FINITE ELEMENT ANALYSIS OF PLANE FRAMES AND TRUSSES

FINITE ELEMENT ANALYSIS OF PLANE FRAMES AND TRUSSES

by
Jack W. Schwalbe, P. E.
Associate Professor of Civil Engineering
Florida Institute of Technology

Robert E. Krieger Publishing Company
Malabar, Florida
1989

Original Edition 1989

Printed and Published by
ROBERT E. KRIEGER PUBLISHING CO., INC.
KRIEGER DRIVE
MALABAR, FLORIDA 32950

Library of Congress Cataloging-in-Publication Data

Schwalbe, Jack W.
    Finite element analysis of plane frames and trusses.

    Includes bibliographies.
    1. Structural frames. 2. Trusses. 3. Finite
element method.   I. Title.
TA660.F7S39 1989   624.1'773   88-9474
ISBN 0-89464-314-2

10   9   8   7   6   5   4   3   2

Errata for Schwalbe, FINITE ELEMENT ANALYSIS OF PLANE FRAMES & TRUSSES

Page	Correction
ix	Line 21, change mathematically to mathematical
ix	Line 34, change analyeds to analysed
3	Line 4, change unnessesary to unnecessary
7	Line 7, change areais to area is
32	In the stiffness matrix, K_{33} should be +
33	In the stiffness matrix, K_{33} should be +
36	In the stiffness matrix, after the boundary conditions are applied, the first term should be +
51	In the matrix K^{-1}, the 2-2 term should be -4
52	In the first equation, -320 should be 320
53	In the first matrix equation, a 1 is missing in the moment vector
54	$M_{f23} = wL^2/12$
56	In the last equation, a close bracket and an = are missing Should be $[]\{\} = \{\}$
56	Change $[K]$ to $|K|$
58	Problem 3-5, change second M_{FBC} to $M_{FCB} = + 108$
62	In the matrix, the 2-2 term should be $2400/(120)^2$
64	In the last equation, the 1-1 term in the stiffness matrix should be k_{ii}
67	Line 11, change transfomation to transformation
67	Change second a_{ij} matrix to a_{ji}
68	a_{78} should read: $$a_{78} = \begin{bmatrix} \cos a_{78} & \sin a_{78} & 0 \\ -\sin a_{78} & \cos a_{78} & 0 \\ 0 & 0 & 1 \end{bmatrix} = \begin{bmatrix} -.894 & -.447 & 0 \\ .447 & -.894 & 0 \\ 0 & 0 & 1 \end{bmatrix}$$
68	a_{87} should read: $$a_{87} = \begin{bmatrix} \cos a_{87} & \sin a_{87} & 0 \\ -\sin a_{87} & \cos a_{87} & 0 \\ 0 & 0 & 1 \end{bmatrix} = \begin{bmatrix} .894 & .447 & 0 \\ -.447 & .894 & 0 \\ 0 & 0 & 1 \end{bmatrix}$$

70	Mid-page, a_{ij}^5 should be a_{ij}^{-1}
71	$S_{ij} = a_{ij} - 1k_{ij} \, a_{ji}$ should read $S_{ij} = a_{ij}^{-1}k_{ij} \, a_{ji}$
71	The 2-2 term of both the last S_{ij} matrix and the last S_{ji} matrix requires the addition of a) after the $\sin^2 a$
71	The next to the last matrix should read S_{ji}
71	In the last matrix, the 2-2 term has an AE/1 that should be AE/L
75	Interchange signs of the 2-3 and the 3-2 terms in S_{12} & S_{21}
75	Change second column of k_{23} to: 0 -2.083 0
75	Change second column of k_{33} to: 0 2.083 0
78	The 3-3 term in J_{ij} should be 4EI/L
88	Note: should read, $a = a_{ij}$
96	Problem 4-12, units of E should read ksi
110	In problem 5-4, it should be noted that the building is square
111	In Problem 5-7, the slab is 12" thick with 6" curbs
120	In the third beam diagram, the shear vector on the right side should be pointing up
122	symmetyry should be symmetry
123	for the member forces due to distributed loading, the moments at joint 2 should be reversed in direction
139	In problem 6-7, change EI = 30 x 10^{6}psi to E = 30 x 10^{6}psi
148	The first term in $J_{22} = .308$
156	Figure for Prob. 7-3, dimension line missing between 4' & 6' and extra 10k symbol should be removed
166	M32 should read: $M_{32} = \begin{vmatrix} K_{11} & K_{13} \\ K_{21} & K_{23} \end{vmatrix}$

To Lyndell my wife,
my love, my life.

Acknowledgment:

Many thanks to the many students who suffered through the proof-reading of the text and problems for this book. There are too many to mention by name, but they know who they are. I hope that they derive some satisfaction to having contributed to this book. I am very grateful.

Jack W. Schwalbe

*   *   *   *   *

A structure has been defined as "any assemblage of material which is intended to sustain loads," and the study of structures is one of the traditional branches of science.

Structures are involved in our lives in so many ways that we cannot really afford to ignore them. Every plant and animal and nearly all of the works of man have to sustain mechanical loads without breaking, and so practically everything is a structure of one kind or another.*

*Gordon, J.E., Structures, Plenum Press, New York and London, 1978

Table of Contents

		Page No.
	Introduction	**ix**
Chapter 1.	**Math Modeling Techniques**	**1**
Chapter 2.	**Introduction to the Stiffness Matrix Method of Structural Analysis**	**28**
	The Bar Element—a Strength of Materials Approach	
Chapter 3.	**The Slope-Deflection Method Reviewed,** SOLUTIONS USING MATRIX ALGEBRA	**45**
Chapter 4.	**The Structural Stiffness Matrix**	**60**
	The Plane Frame Finite Element	60
	Sign Conventions and Coordinate Systems	61
	Coordinate Transformation	65
	Procedure for Forming the Structural Stiffness Matrix	76
	Stiffness Matrix for a Truss Element	78
	The Fixed-Pinned Finite Element	86
	The Fixed-Roller Element	89
Chapter 5.	**The Load and Displacement Vectors**	**98**
	Types of Loading	98
	Load Combinations	101
	Member Forces	108
Chapter 6.	**Analysing Plane Frames and Trusses**	**115**
	Procedure	115
Chapter 7.	**Analysis of Grid Structures**	**144**
	Coordinate Transformation	147
	Shear Center and Torsion	151
	Axial Load Eccentricity	151
	Bi-axial Bending and Unsymmetrical Bending	152

Orientation of Local Axes   153
Super Position   154
Three Dimensional Structures   155

Selected Bibliography   159

Appendices:
A. Selected Review of Matrix Algebra   161
B. Glossary   169
C. List of symbols   171
D. Fixed End Forces and Moments   174

Introduction

The use of computers to solve complex structural analysis problems has, in the past, been restricted to large engineering organizations which could afford the expensive and sophisticated structural analysis programs written for large main frame or mini-computers. Recently, structural analysis software has become available for use on micro-computers. The goal of a structural engineer is to design a structure. Analysis is part of the design process.

With the increasing capability and decreasing cost of micro-computers, individual engineers and small companies as well as colleges and universities may now utilize the speed and economy of computer technology in structural analysis and design. Today, almost every structural engineer is using a computer to perform structural analysis.

These developments make it more important than ever that engineers acquire an understanding of how structures behave under load in order that they may more accurately model the structural problems to be solved and also to better interpret the results of the analyses relative to the real structures. Because of the large volume of data that usually results from a computer analysis, it is important that the engineer develop the ability to judge whether or not the results are realistic.

The purpose of this book is to present some of the techniques of structural mathematically modeling and to introduce the finite element method of structural analysis to the undergraduate engineering student who has successfully completed a traditional, first course in structural analysis.

The use of the term finite element method, in stuctural analysis, is usually reserved for analysis of large, complex, continuums which behave as essentially plate, shell, or solid structures. However, the method used for ordinary framed structures, such as rigid frames and trusses, where the individual members *are* finite elements, may also be used to analyse structures modeled with plate, shell or solid elements.

All structures are three dimensional. However, because of the directionality of the loading and the layout of the structural scheme, many structures may be adequately analyed in two dimensions. For instance, in a typical building, seperate analyses are performed on east-west sections and north-south sections. This book deals with structures which may be adequately modeled as planar structures only.

The matrix method, using the stiffness approach, is the method used in this text. It is an efficient method which may be applied to any type of structure and is the method used in most structural analysis computer programs. While the method is not limited to static analysis, in this book, static loads only are considered.

In finite element analysis, the math model is an assemblage of members, called finite elements, connected at nodes, called joints. These members, or finite elements, having known mechanical properties, may be linear or non-linear; one, two, or three dimensional; elastic or inelastic, and represent the stiffness between joints. In this text, for simplicity, only linear systems composed of one dimensional elements which are prismatic, slender, and made of homogeneous, elastic, and isotropic material will be considered. The student is expected to use consistent units.

In a traditional first course in structural analysis, the student has been introduced to the concept of element stiffness and joint stiffness in the development of the various analytic methods. In this text, these concepts are used in deriving the element stiffness matrix and assembling the structural stiffness matrix.

The material presented in this book was developed from course notes which evolved, over a period of years, for the second of a two course sequence in structural analysis taught at Florida Institute of Technology. Since there did not seem to be a text which completely met the requirements of the course as it was to be presented, it was decided to teach the course from notes. Subsequently, several texts have been published on matrix analysis of structures and on finite element structural analysis. Few of these books present the material in as simple a fashion as was deemed necessary for the level of a typical undergraduate student. Furthermore, none of the books emphasize structural concepts or discuss at any length, mathematical modeling, but instead, dwell on the method.

The author believes this book is unique in two ways. First, Chapter (1) brings together in one place, for the first time, some of the techniques of math modeling that lie scattered throughout the literature. Second, the simplicity and elegance of the matrix method of structural analysis are carefully explained and illustrated. As will be shown later, building the structural stiffness matrix is essentially a bookkeeping procedure which starts with the information contained in the idealized structure. The structural stiffness matrix together with the load and displacement vectors comprise the math model which represents the idealized problem to be solved.

The attempt has been made in this book to present the material in a way that will lead the student to an understanding of the way structures behave under load and not simply to present another structural analysis technique. With this in mind, it was felt that a chapter on math modeling techniques was essential. The aim, always, is to help the student attain an understanding of the way stuctures respond to loads.

In structural analysis, the concern is usually to determine the displacements and/or stresses resulting from a given set of loading conditions. The basic equation to be set up and solved is $P = KX$, where P represents the vector of external forces applied to the structure; K represents the structural stiffness matrix; and X represents the solution vector or the joint displacement vector. The method of solution used herein to accomplish these ends is the stiffness method, also called the displacement method.

In the stiffness method, the element stiffness matrices are formed by imposing unit displacements on the nodes of the elements individually and calculating the resulting element forces. The individual element stiffness matrices are then assembled into the structural stiffness matrix.

The structural analysis procedure is:

I. STRUCTURAL IDEALIZATION:
  1. Represent the idealized structure by an assemblage of elements and joints or nodes.
  2. Number all joints and members.
  3. Locate the joints (nodes) in space by determining their coordinates in a global coordinate system.
  4. Establish the element connectivity.
  5. Assign geometric and material properties to the elements.
  6. Determine the boundary conditions and loads.

II. BUILDING MATH MODEL:
  1. Form the element stiffness matrix.
  2. Rotate the element stiffness matrices into the global coordinate system.
  3. Assemble the element stiffness matrices to form the structural stiffness matrix.
  4. Apply the boundary conditions and loads.

III. SOLUTION:
  1. Solve for the unknown joint displacements.
  2. Compute the element forces from the joint displacements.
  3. Compute the reaction forces at the supports.
  4. Perform an equilibrium check.

5. Analyse the results with respect to the real structure.

The student is not expected to write any programs for this course nor are there any computer programs presented herein. However, the method as developed, is easily programmed should the need arise. Structural analysis computer software is readily available on the open market, affordable and, more importantly, reliable. The material presented in this text is general in nature, is applicable to most structural analysis computer programs, and is intended to facilitate the interpretation of the user's manuals.

A working knowledge of matrix algebra is necessary in order to carry out the numerical solutions to the problems using the matrix method of structural analysis. A selected review of matrix algebra is presented in Appendix A. The solution to some of the example problems includes the matrix algebra in detail, as a further aid to the student.

The student should master the method before relying on a structural analysis computer program, for reasons previously stated.

For problems with three or more degrees of freedom, it is recommended that the student have available a program for inverting the stiffness matrix and performing the matrix multiplication. This will allow the student to solve many more problems as well as larger problems. These programs are readily available for programmable calculators and microcomputers.

It must be emphasized that the student is to develop judgement in setting up the problem to be solved (preparing the math model), and leaving the number crunching to the machine, which can do it faster and more accurately. But, while the machine gives instant answers, it does not tell the student how it got those answers or what the answers mean in terms of the real structure. Therefore, the results of the analyses must be interpreted as to their significance to the real structure.

CHAPTER I

MATH MODELING TECHNIQUES

Many structural analysis computer program user's manuals state that the assumption is made that the user of a finite element program is knowledgeable in finite element theory and applications. However, it is the user's responsibility to develop a mathematical model and to properly present the input data for that model to the structural analysis program being used. Since the results of an analysis are only as good as the input data, the modeling will greatly influence the results.

At the present time, there are no textbooks available which explain the methodology required to take a complex structure and mathematically represent it for analysis. This is probably because math modeling requires a great deal of judgement, which is best gained by experience, and is difficult to learn from books. Nevertheless, there are certain principles that can be explained in a textbook, and some of these are presented in this chapter and will be expanded upon and illustrated by example in subsequent chapters of this book.

THE MODELING PROCESS

Before one can analyse a structure, a mathematical model of that structure must be prepared. (A structural math model has been defined as the idealized representation of the structure using basic mechanical and/or structural elements. It is often referred to as the structural idealization) in texts dealing with finite element analysis of structures.

The process of creating a math model proceeds from the structural drawings, which are an interpretation of the real structure, to an idealization of that structure as shown on the drawings, to a mathematical representation of the idealized structure that lends itself to analysis (Fig.1-1). Proper construction of a model will lead to a mathematical analysis which will produce results which accurately represent the response of the actual structure to the actual applied loads and constraints and, also, minimize the time required for solution.

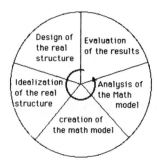

FIG. 1-1

In the initial stage, it is often good practice to construct a coarse model of the simplest elements which will produce the desired information. This coarse model is not intended to reflect the true structural geometry in great detail, but instead it should give an indication of the overall structural response to the applied loads. Since this model is only intended to be a simplified representation of the actual structure, the analyst should deliberately overlook small details or discontinuities in the structure. The minimum number of nodes that will characterize the actual response should be used.

The coarse model is only intended to provide overall structural deformations and internal force distributions. From these deflections and forces, relative stresses can be approximated for large areas of the structure.

This gross deflection and stress information is extremely useful in making many engineering decisions. This crude analysis may show such things as:

- stability
- load paths
- effect of adding or subtracting members
- effect of adding or subtracting loads or constraints
- points of high or low stress
- locations of large displacements

Thus, although the results of this crude analysis may not produce very accurate results, they are invaluable as an engineering aid to guide the analyst to a better understanding of the gross structural behavior.

Once the coarse model has indicated the overall structural characteristics, more detailed models may be developed for analysis in areas where stresses and deflections must be more precisely determined. Here again, the engineer is called upon to exercise judgement. The aim is to create a math model which is simple/economical and, at the same time, produces the desired results. The use of unnessesary nodes increases the amount of storage required and hence time and cost. On the other hand, the coarser models generally indicate stiffer structures, ie., the more elements used, the more flexible the structure appears.

In some cases, the results of the analysis of the coarse model will be found to be sufficient and further analysis unnecessary. Figure(1-2), for instance, shows a detail of a typical beam-to-column connection such as might be used at joint (3) of Figure(1-3). The details of the connection would not be included in the math model representing the idealized structure, but the analysis of the coarse model would yield the necessary forces and moments to design the connection in detail. On the other hand, the degree of fixity assumed for the joints of the structure may have a significant affect on the resulting internal forces and displacements.

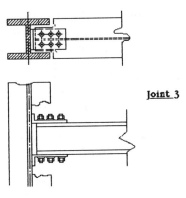

Joint 3

Fɪɢ. 1-2

TECHNIQUES OF STRUCTURAL IDEALIZATION

From the set of drawings representing the real structure, a line drawing is prepared, Fig.(1-3), where each line represents the center of gravity axis of a member. This is the idealized structure and the process is sometimes referred to as "creating the mesh". This line drawing is plotted with respect to a set of reference axes called the structural or "global" coordinate system and the nodes and elements are numbered for identification. The structural coordinate system used in this text is a right hand system with the X-axis horizontal to the right and the Y-axis vertically up, Fig. (1-3). Turning X into Y advances a right hand thread out of the x-y plane establishing the positive Z direction.

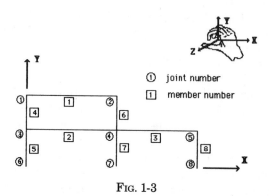

FIG. 1-3

NODES OR JOINTS

Nodes are the points selected on the structure and within the math model to describe the basic layout or shape of the structure. A node or joint is required wherever two or more members meet, where loads or boundary conditions are to be applied, or where output is desired.
Nodal information includes:

- position in space relative to the Structural Coordinate System (nodal coordinates)
- degrees of freedom relative to the Structural Coordinate System
- connectivity to other nodes via elements

(An important consideration is the order of the node numbering. Nodal numbering should be done in such a manner as to minimize the numerical difference between any two nodes which are connected by an element. This will minimize the bandwidth of the stiffness matrix and lead to a more efficient solution.) Some computer programs provide bandwidth minimization routines so that the nodes may be numbered arbitrarily, however, nodal numbering considerations may still save time, even with a minimization routine.

ELEMENTS

Elements are the mathematical representation of the structure between nodes. This mathematical representation is in the form of a stiffness matrix. Elements are described by the following information:

- element type
- material properties
- geometrical configuration and dimensions
- section properties
- location in space by nodal connectivity

When constructing the idealized model, one has to choose the correct type of elements to represent the structural behavior. The use of complex elements where simple ones will do will waste time and money thru increased demand for core storage. The range of element types varies with each particular computer program, but the following types are most common.

Discrete elements, such as bar or beam elements, are elements which when used singularly, will accurately represent a structural member of the same type. One beam element is normally sufficient to model a beam. However, a larger number of plate elements is necessary to obtain accurate stress data when modeling a plate. Discrete elements are often referred to as one dimensional elements since they have length in one direction only.

Finite elements of continua, such as plate , shell, or solid elements are used to model structures which behave structurally as plates (primarily two way flexure), shells (primarily membrane action) , or solid structures (in which the stresses vary significantly through the thickness). Beam elements are sometimes used to represent a continuum for preliminary investigation of the displacement pattern and areas of high stress gradients, and to aid in the establishment of the boundary conditions and loading conditions.

In this text, for simplicity, only structures which may be modeled with bar, grid, and/or beam-column elements will be considered. This may seem restrictive, however, the purpose herein is to impart an understanding of the finite element method of analysis, which is essentially independent of the type of structure or finite element. The method, including most modeling principles and matrix analysis techniques, is essentially the same for continuum type structures as it is for discrete element type structures.

(Bar elements are one dimensional elements (having length in only one direction) which are used to represent members which may only carry axial forces, ie., forces directed along the members longitudinal axis. The member is assumed to be pin-ended with only translational degrees of freedom at the nodes. This type of element is used to represent members of a truss and is referred to, in this book, as a truss element.)

(Beam-column elements are one dimensional elements which are used to represent members which may carry shear forces, bending and twisting moments, as well as axial forces. In this book these elements are called plane frame elements. The member may have any end restraints desired and may have three translational and three rotational degrees of freedom at each node. This element is used to represent members in continuous beams and rigid frames.)

Grid elements are used to model planar structures which carry out of plane loads.

It is very important to be familiar with the formulation of the element stiffness matrix for the particular element to be used in each application. For instance, were shear stiffness and displacement considered so that one element may represent a shear panel; can the element handle unsymmetric bending, loads not through the shear center, eccentric axial loads, etc.

When two or more members intersect, but their center of gravity axes do not, the common end of their axes must be connected such that each end of each element will see the same displacement. One way to accomplish this is to connect each member to a common node by imaginary members known as stiff arms, such as shown in Fig. (1-4). They are called stiff arms because they are assigned a stiffness many times that of the members to which they are connected.

A word of caution however, if the member stiffnesses differ by a factor greater than about 10,000, numerical problems may result in the stiffness matrix. Some computer programs treat this problem by using beams in series thereby eliminating a joint. In more sophisticated programs, nodes are placed at the common ends of each member and the two nodes are coupled mathematically instead of using fictitious members. This aspect of math modeling is beyond the scope of this text.

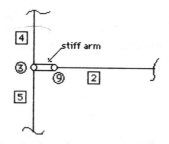

FIG. 1-4. *Structural idealization of joint (3) of Figures (1-2) and (1-3).*

PHYSICAL AND MECHANICAL PROPERTIES

For each and every element in the model, values must be assigned for the cross-sectional properties and material properties necessary to compute the terms of the element stiffness matrix and, if required, stresses.

The physical properties required for calculating element stiffnesses depend on the type of element. For the two-dimensional beam-column element, for instance, the physical properties required are the cross-sectional area and the moment of inertia about the axis of bending, while for the truss element only the cross-sectional area is needed. For stress analysis, additional information is required, again depending on the type of element.

The material properties required for stiffness calculations are also element dependent. For the two-dimensional beam-column element, the only required material property is the modulus of elasticity.

The user's manual for each particular computer program will specify the required physical and material properties for each element.

BOUNDARY CONDITIONS AND MEMBER RELEASES

Basic decisions must be made at this point in the process as to whether or not the structure may be adequately represented by a planar model or must a three dimensional model be used and whether the structure is a truss (no joint moments) or a frame (full joint moment) or something in between. In addition to the degree to which the joints can support forces, decisions must be made as to whether or not the individual members can support particular forces. In the event that a member cannot support a force in a particular direction, the degree of freedom associated with that direction must be released at one end of the member. For instance, in the expansion joint, joint (10) of Figure (1-5), member (9) cannot support a longitudinal force. Therefore, the end of member (9) is allowed to freely translate in the longitudinal direction. This is accomplished mathematically by altering the member stiffness matrix, refer to Chapter (6).

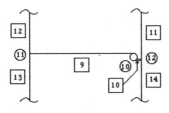

Fig. 1-5

The boundary conditions or constraints and the external forces or loads are then applied to the line drawing Fig.(1-6). The real structural problem should be studied to determine the load distribution and the degree of constraint to be used in the math model. There are no such things as concentrated loads, perfect fixity or perfectly pinned connections, for instance. The type of loading has a great affect on the math model. Whether the loads are concentrated or distributed, and if concentrated, where they are located on the structure affects the nodal locations and element types.

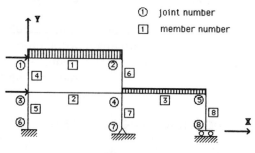

FIG. 1-6

Boundary conditions are known values of force, displacement, temperature, etc. applied at selected nodes of the structure. The more common boundary conditions, for a planar structure, with which the student should already be familiar, are shown in Figure (1-7). (a) indicates a support which may not translate in the direction perpendicular to the line of rolling, but may rotate therefore there is only one unknown force.

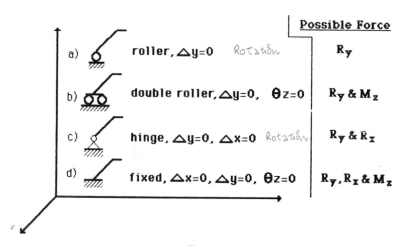

FIG. 1-7

(b) indicates by the double roller that no displacement perpendicular to the line of rolling may occur, but a moment may exist at the joint, precluding any rotation as well. In this case, there is an unknown force and an unknown moment. A hinge, as shown in (c) indicates that no translations may exist, but rotation is allowed, therefore, there are two unknown forces. (d) is a fixed joint, which means no translations or rotations are allowed. All three forces are unknown. There are simlar boundary conditions for three dimensional structures.

It is most likely that in a real structure the boundary conditions are not perfectly pinned or fixed, but something in-between. In addition, there are many other constraints which may be used for various special purposes, such as those shown in Figure (1-8).

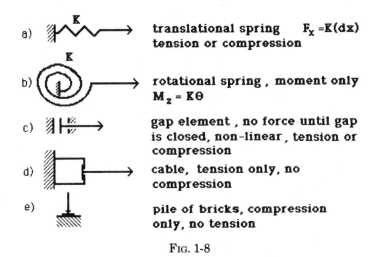

a) translational spring $F_x = K(dx)$ tension or compression

b) rotational spring , moment only $M_z = K\theta$

c) gap element , no force until gap is closed, non-linear , tension or compression

d) cable, tension only, no compression

e) pile of bricks, compression only, no tension

FIG. 1-8

(When only one member is attached to a support joint such as in Figure (1-7), the release of a restraint should be accomplished by the use of a boundary condition and not by releasing the restraint on the member. This is a common modeling error, which is to be avoided.) Specifying too many releases for member forces when modeling hinges or rollers can lead to structural instability or a singularity in the stiffness matrix. For instance, releasing the same degree of freedom for all members joining at a node will result in a zero stiffness term for the joint, and therefore, a singularity.

SUBSTRUCTURING

A modeling technique called substructuring may be used to increase the efficiency of the solution and may allow the solution of problems where the stiffness matrix is too large for practical considerations. This technique is particularly useful when using a structural analysis computer program on a computer which has insufficient storage to handle a large stiffness matrix. This is becoming a more common problem as more structural programs are being written for use on micro-computers using the matrix method. Since the matrix method requires the storage of the structural stiffness matrix, the size of the problem which may be solved is limited by the amount of memory available.

In using substructuring, the structure is subdivided into sub structures, each of which is described by a matrix set of force displacement equations and analysed separately with the appropriate loads and constraints. Then the results are combined, forcing compatibility and equilibrium at the common boundaries. The structure is subdivided into the minimum number of substructures necessary, as the fewer substructures created and the fewer common boundary joints, the more efficient the solution.

⊛ (When a structure contains an axis of symmetry or anti-symmetry and if the loads are also symmetrical or anti-symmetrical about the same axis, the structure may be separated through the axis of symmetry or anti-symmetry into substructures, only one of which need be analysed. A savings in the time required for both modeling and analysis will result from modeling only the structure on one side of the axis. Specific boundary conditions must be imposed at the cut points such that no consideration need be given to the other half of the structure.)

(In a symmetric structure, one half of the structure is a mirror image of the other half in terms of geometry, distribution of material properties, magnitude and placement of loads, and type and arrangement of constraints. Then, the displaced shape (elastic curve) of the structure will be symmetric about an axis of symmetry and , in addition, the values of the internal forces, reactions, and displacements will also be symmetric.) See Figure (1-9).

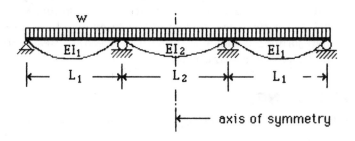

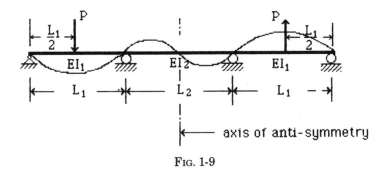

FIG. 1-9

When the structure is symmetrical, but the loads are anti-symmetrical, the displaced shape, the internal forces, and the reactions will be anti-symmetrical.

When the structure is symmetrical, but the loads are neither symmetrical nor anti-symmetrical, the loads may be resolved into an equivalent set of symmetrical and anti-symmetrical loads acting separately. The half-structure is then analysed twice, once with the symmetrical loading and boundary conditions and once with the anti-symmetrical loading and boundary conditions, and the results combined. This may seem to increase the analysis time, but actually, it will shorten it since only one-half the structure will be analysed. The symmetrical and anti-symmetrical loads and boundary conditions are shown in Figure (1-10) for the structure of Figure (1-9).

To obtain the results for the whole structure, the results of the two analyses are superposed.

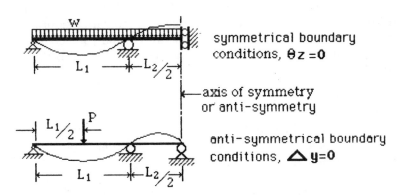

FIG. 1-10

EVALUATION OF RESULTS

When the mathematical solution has been completed, and the results interpreted in terms of the math model, the modeling process must be reversed and conclusions drawn relative to the real structure. The results must be examined in light of the assumptions that were made in the idealization of the structure and in the modeling process.

When the student has mastered the matrix method and has aquired a good understanding of the principles presented in this book, the next step is to learn to use a structural analysis computer program. The use of a computer to perform the tedious calculations frees up the analyst to spend more time on more challenging and interesting part of the process; those phases of the analysis process requiring the use of judgement. These are the preparation of the math model and the evaluation of the results in regards to the real structure. The analyst may then have time to play "what if" games, i.e., to investigate other loading conditions and/or the effects of varying certain parameters such as member stiffness, boundary conditions, additional bracing, etc.

Since most structural analysis computer programs use a variation of the method as presented herein, an understanding of this method should lead to a greater awareness of just what the computer is doing and allow the analyst to prepare a mathematical model which is more representative of the real structure and to better understand and interpret the results.

CHECKING

A modeling tool which may be used at this point is plotting. With all the proper values for joint coordinates, member type, loads, and constraints, a plot is an excellent way to see if there are any errors in the geometry of the structural idealization. Modeling errors are expensive in time and money if not discovered and corrected prior to carrying out the analysis. Some common errors which will show up in a plot are:

- missing element
- missing node
- incorrect nodal coordinate
- incorrect node or element numbering (repeat or skip)
- Incorrect element connectivity

Depending upon the relative sophistication of the plotting software, additional errors which may be revealed by plotting are:

- incorrect boundary conditions
- incorrect member releases
- incorrect loading

MATHEMATICAL MODEL

When the analyst is satisfied with the idealized depiction of the structure, a mathematical representation is prepared. In the matrix method of stuctural analysis, the mathematical representation of the structure is called the structural stiffness matrix. A substantial portion of this book is devoted to building the structural stiffness matrix from the idealized structure.

The three basic tools of structural analysis, which will be used , are:

1. The equations of static equilibrium, i.e., the summation of forces and moments at all points and in all directions are equal to zero.
2. The stress-strain relationships, assuming that the material properties are linear, elastic, homogeneous, isotropic, constant with time and temperature, and are the same in tension and compression.
3. Compatibility equations, which state that the members framing into a rigid joint must undergo the same displacements, the members must remain attached to the joints, and the boundary conditions must be satisfied.

COMPUTER OUTPUT

The selection of an appropriate math model is influenced by the type of results desired. The nodal locations, element types, boundary conditions, member releases, load distribution, and displacement data are all affected by the type and location of output data required. What displacements are desired and where? Are internal forces required in all areas of the structure or only in a limited number of locations? Is stress output desired? Answers to these types of questions will affect which physical and mechanical properties are necessary as input data.

There are many different types of information which can be obtained from many different types of computer programs and many different forms in which the output may be presented. The engineer is often called upon to be familiar with more than one structural analysis progam each of which may perform a unique task. There are special purpose programs and general purpose programs. A special purpose program might be one which only analyses plane frames and trusses while a general purpose program might be able to handle three-dimensional structures and contain a large library of elements.

Example Problem 1-1. Prepare a structural idealization of the steel structure of Figure (1-11). The end of the center of gravity axis of the w8x24 which is in the concrete wall is to be taken as the origin of coordinates. The elevation at that point is + 6.0 feet. The lower end of the channel is at elevation + 3.0 feet. The w8x24 is 5 feet long, the angle is 4 feet long, and the channel is 3.75 feet long. There is a concentrated load of 5000# directed vertically downward, at the intersection of the members and a uniformly distributed load of 0.2 K/in directed vertically downward, along the longitudinal axis of the w8x24. Discuss the boundary conditions.

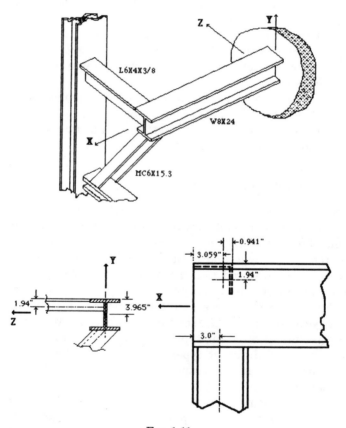

Fɪɢ. 1-11

Solution:

Figure (1-12) shows a sketch of a structural idealization which has been chosen to represent the structure of Figure (1-11), according to an interpretation of the problem as presented. All quantities have been specified in consistent units, ie.; kips and inches.

Along with the material and cross-sectional properties of the members, the boundary conditions are all that is necessary to proceed with the creation of the math model. The material and cross-sectional properties may be determined from the appropriate specification.

If an analysis of the support structure has been performed, the forces and displacements at joints (1), (3), & (4) may be included with the input data of this solution as part of the boundary conditions. However, if this information is not available, the boundary conditions to be used are a matter of judgement. The degree to which the members may displace at joints (1), (3), & (4) may be determined by including the support structure in the analysis, which will increase the size of the stiffness matrix. It may be more economical, and perhaps sufficiently accurate, in this case, to analyse the simple model twice; once with fixed boundaries and once with pinned boundaries. In the general case, however, there may be many combinations possible.

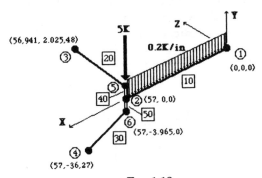

FIG. 1-12

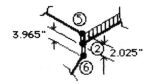

FIG. 1-12A

Note that there are intermediate nodes used at one end of members (20) and (30), with stiff arms connecting them to joint (2). This is because the center of gravity axes of the three members are not congruent, Fig. (1-11). If the solution method could handle series beams, nodes (5) and (6) would become unnecessary or if the solution method utilized nodal coupling, members (40) and (50) would become unnecessary.

It may be, for design purposes, only the w8x24 is of interest. Since the main structure and the loads lie in the X-Y plane and there is not likely to be significant translation in the Z direction, the structure may be modeled in two dimensions making appropriate assumptions regarding the boundary conditions, if, in the opinion of the analyst, torsion is not a problem.

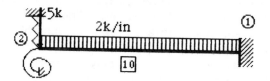

FIG. 1-13. *Two-dimensional Model.*

Example Problem 1-2. Figure (1-14) shows a braced frame structure which is symmetrical about the centerline and subjected to a uniform gravity load of 2 k/ft. Draw the structural idealization, number the members and joints, and indicate the boundary conditions.

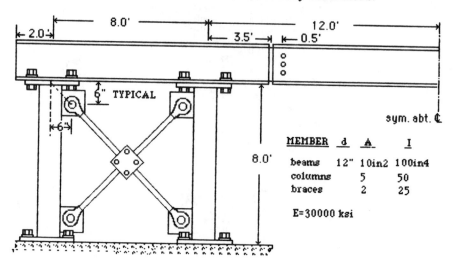

MEMBER	d	A	I
beams	12"	10in2	100in4
columns	5		50
braces	2		25

E=30000 ksi

FIG. 1-14

Solution:

The idealized structure is shown in Figure (1-15) with the members and joints numbered. Notice where there are no joints! The overhanging portion of the beam may be replaced as shown in Figure (1-16). There is no joint necessary at the intersection of the diagonal bracing members as output is not usually required there.

All bolted joints in this structure are considered incapable of transmitting a moment and are therefore pinned joints. This should be accomplished at joints (11) and (12) by pinned boundary conditions and not by releasing the rotational stiffness of both members (1) and (2). Since the diagonal bracing members are bar elements, they will provide no rotational stiffness to the joints, therefore, members (1) and (2) are the only source of rotational stiffness for joints (11) and (12).

The symmetrical boundary conditions applied to joint (10), for this planar structure, are that the joint may translate in the vertical direction only, no rotation is allowed.

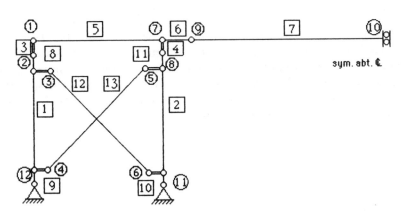

Fɪɢ. 1-15

$$P = 2K/ft\ (2\,ft) = 4\ K \downarrow$$
$$M = \frac{2\,K/ft\ (2\,ft)^2}{2} = 4\ K\text{-}ft \curvearrowright$$

Fɪɢ. 1-16

Example Problem 1-3. Show on separate sketches the symmetrical and anti-symmetrical idealizations with the appropriate loadings and boundary conditions for the structure of Figure (1-17).

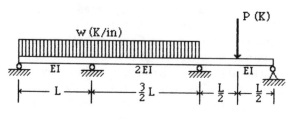

FIG. 1-17

Solution:

The structural idealization need only be done once but two sub- – structures must be created. Each substructure is analysed with the appropriate boundary conditions and loads, as shown in Figure (1-18). The sum of the results of the separate analyses will yield the solution. The analyses may be carried out on either the two left-half (a&c) or the two right-half (b&d) substructures.

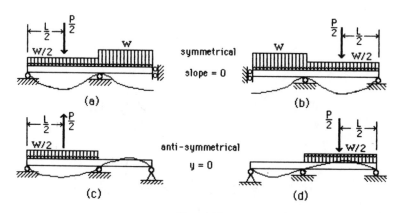

FIG. 1-18

Example Problem 1-4: The structure shown is symmetric. The loading shown is non-symmetric. Show the structural idealization utilizing symmetry.

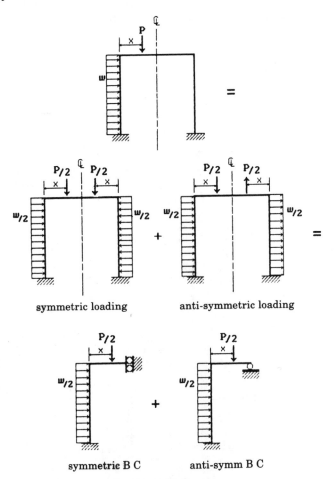

symmetric loading   anti-symmetric loading

symmetric B C   anti-symm B C

Equivalent half structures

PROBLEMS FOR SOLUTION

I-I. Prepare a sketch of the structural idealization of the lifting beam shown in Figure (1-19). Discuss the effect on the efficiency of the solution of the number of elements used in this case.

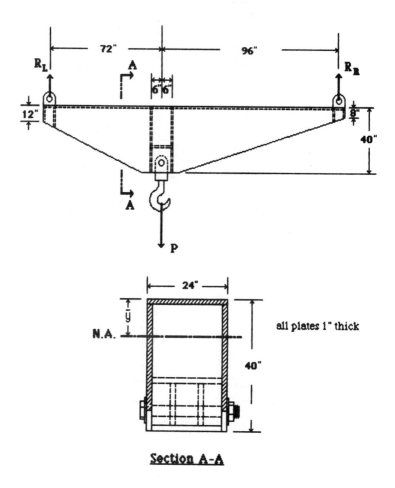

Fɪɢ. 1-19

1-2. Indicate with sketches how you would idealize the structure shown in Figure (1-20), using the appropriate loadings and boundary conditions.

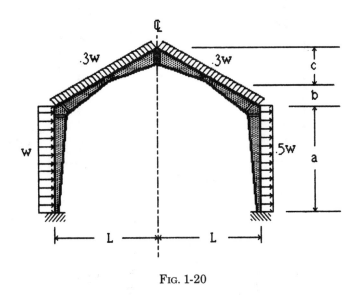

FIG. 1-20

1-3. Figures (1-21) and (1-22) show a crane support structure. Prepare a structural idealization in three dimensions making use of symmetry where possible.

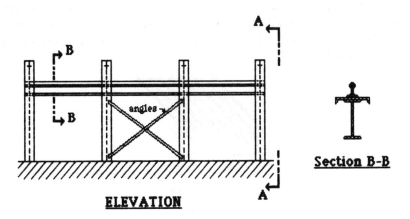

Section B-B

ELEVATION

FIG. 1-21

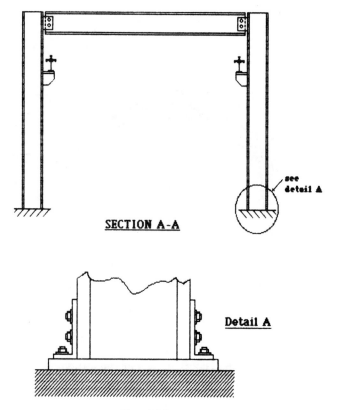

SECTION A-A

Detail A

Fig. 1-22

I-4. Discuss how you would idealize the structure shown in Figure (1-23) assuming you wish to minimize storage space in the computer. The members supporting the roof beams are Vierendeel Trusses as shown in Figure (1-24).

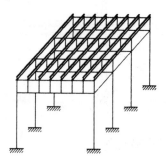

Fig. 1-23

Vierendeel Truss

FIG. 1-24

I-5. The St. Louis Arch is subjected to wind loads as well as a very large dead load. Discuss any differences in your recommended math models for analyzing for these two loading conditions taken separately.

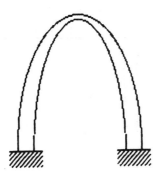

FIG. 1-25

l-6. Sketch the structural idealization of the structure of Fig. (1-26).

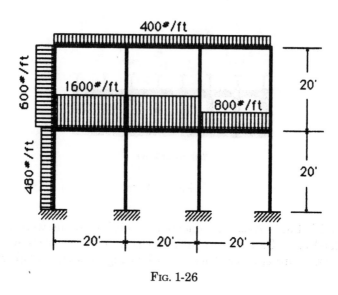

FIG. 1-26

l-7. Sketch the structural idealization of the structure of Fig. (1-27).

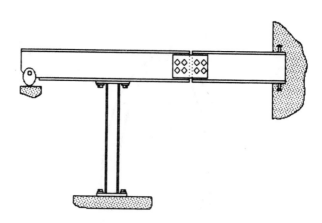

FIG. 1-27

1-8. Sketch a structural idealization of the cantilever bridge shown in Fig. (1-28).

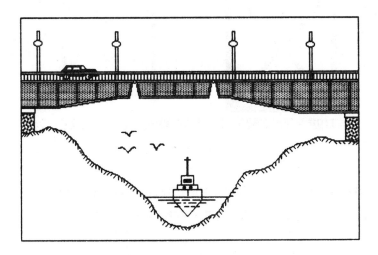

FIG. 1-28

1-9. Sketch the structural idealization of the truss and girder bridge shown in Fig. (1-29).

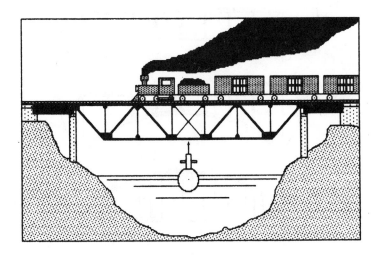

FIG. 1-29

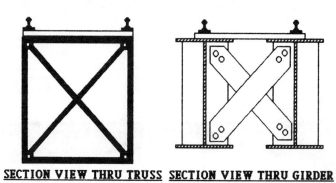

SECTION VIEW THRU TRUSS   SECTION VIEW THRU GIRDER

Fɪɢ. 1-30

I-I0. Discuss the load paths for the bridge structures shown.

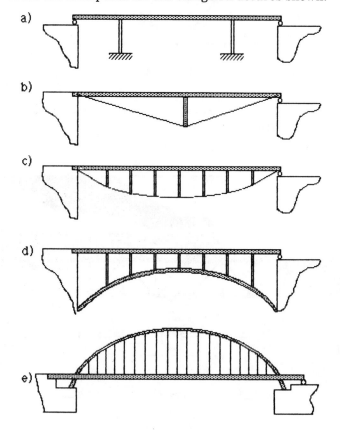

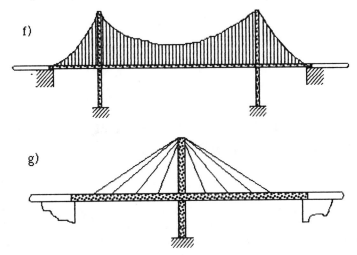

I-II. Select a structure located nearby and prepare:

1. A sketch of the structure
2. A structural idealization

CHAPTER 2

INTRODUCTION TO THE MATRIX METHOD OF STRUCTURAL ANALYSIS

In the Matrix Method of Structural Analysis, as used in this text, the matrix equation is established using the Stiffness Method (Displacement Method). The development using the Flexibility Method (Force Method) is presented in many of the references.

The equation to be established is the force equilibrium equation, $P = K X$. Where K represents the stiffness of a structure, P represents the external forces acting on that structure, and X represents the displacements of the joints of the structure.

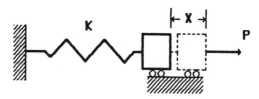

Fig. 2-1

The problem is stated graphically in Figure 2-1. This is the idealized structure for a simple tension or compression member (if buckling is not considered). This most simple of structures is used to illustrate the basic method which will, with slight modification, be used to analyse more complex structures.

THE BAR ELEMENT, A STRENGTH OF MATERIALS APPROACH

A link (truss member, rod, bar, etc.) is any element which is pin-ended such that it may carry axial load only, either tension or compression.

If the ends of the element are given a relative displacement, the force is known to be directed along the longitudinal axis of the element and only the magnitude of the force need be determined.

$$\text{Fig. 2-2}$$

The magnitude of the force developed is equal to the stiffness of the element times the relative displacement of the two ends.

$$P = K(X_i - X_j)$$

From a strength of materials approach, using Hook's Law:

$$\text{stress/strain} = E$$

$$\text{stress} = P/A, \text{ strain} = (X_i - X_j)/L$$

$$(P/A)[L/(X_i - X_j)] = E$$

$$P = [AE/L][(X_i - X_j)], \text{ or } P = K(X_i - X_j)$$

$$K = P/(X_i - X_j) \text{ or } K = AE/L$$

which represents the stiffness of the element between joints (i) and (j).

If joints (i) and (j) are given a relative displacement, $(X_i - X_j)$, the following forces result:

$$P_i = K(X_i - X_j) = AE/L(X_i - X_j)$$

$$P_j = -K(X_i - X_j) = -AE/L(X_i - X_j)$$

$$P_i = -P_j = AE/L(X_i - X_j) = AE/L(X_i) - AE/L(X_j)$$

$$P_i = (AE/L)(X_i) - (AE/L)(X_j)$$

$$P_j = -(AE/L)(X_i) + (AE/L)(X_j)$$

In the development of the moment distribution method, the student was introduced to the concept of locking and unlocking joints. That concept will now be used to illustrate how the structural stiffness matrix is assembled in the matrix method of structural analysis.

Consider the structure shown in Figure (2-3). Joints (1) and (3) are fixed against all possible displacements. Joint (2) may translate in the **X** direction only. If a force is applied in the **X** direction at joint (2), joint (2) translates in the **X** direction causing member (12) to be in tension and member (23) to be in compression.

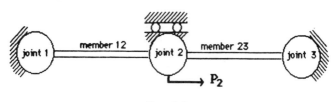

FIG. 2-3

The **laws of equilibrium** require that the sum of the member forces acting on joint (2) must equal the load applied to joint (2).

The **laws of compatibility** require that the ends of the members connected to joint (2) experience the same displacement as joint (2).

The **elasticity relationships** will be used to establish the matrix equation, **P** = **KX**, by locking and unlocking joints.

The procedure is as follows:
Remove all loads.
Lock all joints. (Note that when a joint is locked, no forces may pass through it!)
Unlock joint (1) and give it a displacement in the + **X** direction, Fig 2-4.
Calculate the member forces, **P**, developed at the joints.

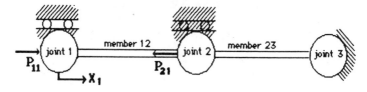

FIG. 2-4

Using double subscript notation; the first subscript refers to the joint at which the force occurs, the second refers to the joint which is given the displacement.

$$P_{11} = K_{12}X_1$$

This equation reads: The force at joint (1) due to a displacement at joint (1) is equal to the stiffness of member (12) times the displacement of joint (1). Note that K_{12} equals the stiffness between joints (1) and (2) and K_{23} equals the stiffness between joints (2) and (3) and, according to Maxwell's Reciprocal Theorum, $K_{12} = K_{21}$ and $K_{23} = K_{32}$.

So, due to the imposed displacement of joint (1) in the $+X$ direction, a force P_{11} is developed at joint (1), in the $+X$ direction, equal to $K_{12}X_1$ and concurrently a force P_{21} equal and opposite to P_{11}, is developed at joint (2) equal to $-K_{12}X_1$.

Lock all joints.

Unlock joint (2) and give it a displacement in the $+X$ direction [Fig.(2-5)]. Two forces will be developed at joint (2). A force in member (12) and a force in member (23). Concurrently, forces will be developed at joints (1) , and (3).

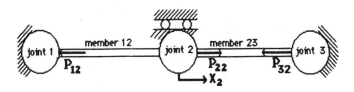

FIG. 2-5

$$P_{22} = K_{12}X_2 + K_{23}X_2$$
$$P_{12} = -K_{12}X_2$$
$$P_{32} = -K_{23}X_2$$

Lock all joints.

Unlock joint (3) and give it a displacement in the $+X$ direction [Fig.(2-6)]. The following forces result:

$$P_{33} = K_{23}X_3$$
$$P_{23} = -K_{23}X_3$$

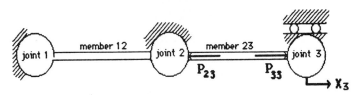

FIG. 2-6

The total force at each joint is*:

$$P_1 = P_{11} + P_{12} = K_{12} X_1 - K_{12} X_2$$

$$P_2 = P_{21} + P_{22} + P_{23} = -K_{12} X_1 + K_{12} X_2 + K_{23} X_2 - K_{23} X_3$$

$$P_3 = P_{32} + P_{33} = -K_{23} X_2 + K_{23} X_3$$

or

$$P_1 = K_{12} X_1 - K_{12} X_2$$

$$P_2 = -K_{12} X_1 (K_{12} + K_{23}) X_2 - K_{23} X_3$$

$$P_3 = -K_{23} X_2 + K_{23} X_3$$

If the substitution is made; $K_{22} = K_{21} + K_{23}$, which is read:

The stiffness of joint (2) is equal to the force developed at joint (2) due to an imposed displacement at joint (2), or, K_{22} *equals the sum of the stiffness of the members attached to that joint.* This concept is similar to that developed for moments in the moment distribution method i.e., the joint stiffness is equal to the sum of the stiffnesses of the members connected to the joint. Then ,by similar reasoning, in this case, the stiffness of joint 1 is $K_{11} = K_{12}$ and the stiffness of joint 3 is $K_{33} = K_{23}$.

Rewriting in matrix form : $P = KX$

$$\begin{Bmatrix} P_1 \\ P_2 \\ P_3 \end{Bmatrix} = \begin{bmatrix} K_{11} - K_{12} & & \\ -K_{21} & K_{22} - K_{23} & \\ & -K_{32} & -K_{33} \end{bmatrix} \begin{Bmatrix} X_1 \\ X_2 \\ X_3 \end{Bmatrix}$$

Now the boundary conditions or constraints are applied. Since joints (1) and (3) are fixed, their displacements must be zero. Then all terms which contain X_1 or X_3 must be eliminated from the equations. This is accomplished by crossing out the appropriate rows and columns from the stiffness matrix.

$$P_1 = K_{11} X_1 - K_{12} X_2$$
$$P_2 = -K_{21} X_1 + K_{22} X_2 - K_{23} X_3$$
$$P_3 = -K_{23} X_2 + K_{33} X_3$$

$$X_1 = X_3 = 0$$

* The force equilibrium equations

$$\begin{Bmatrix} P_1 \\ P_2 \\ P_3 \end{Bmatrix} = \begin{bmatrix} K_{11} & -K_{12} & \\ -K_{21} & K_{22} & -K_{23} \\ & -K_{32} & -K_{33} \end{bmatrix} \begin{Bmatrix} X_1 \\ X_2 \\ X_3 \end{Bmatrix}$$

Which yields the equation:

$$P_2 = K_{22}X_2$$

the solution of which is:

$$X_2 = K_{22}{}^{-1}P_2$$

Now that the displacement of joint (2) is known, the forces at joints (1) and (3) may be determined from the relationships:

$$P_1 = -K_{12}X_2, \quad \& \quad P_3 = -K_{23}X_2$$

EXAMPLE PROBLEM 2-1:

Using the matrix method, determine the stresses in the stepped bar loaded as shown in Figure (2-7). Joint (1) is fixed against all displacements, joints (2) and (3) are free to translate in the X direction only. The material is steel with a modulus of elasticity, E , of 30,000 ksi.

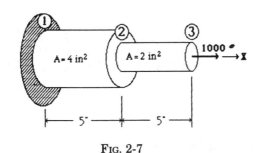

FIG. 2-7

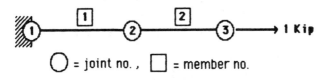

O = joint no. , ☐ = member no.

FIG. 2-8. *The Idealized Structure.*

THE STIFFNESS MATRIX:

$$K = \begin{bmatrix} K_{11} & K_{12} & K_{13} \\ K_{21} & K_{22} & K_{23} \\ K_{31} & K_{32} & K_{33} \end{bmatrix} = \begin{bmatrix} P_1/X_1 & P_1/X_2 & P_1/X_3 \\ P_2/X_1 & P_2/X_2 & P_2/X_3 \\ P_3/X_1 & P_3/X_2 & P_3/X_3 \end{bmatrix}$$

Column 1 = joint forces due to X_1, all other displacements prevented.
Column 2 = joint forces due to X_2, all other displacements prevented.
Column 3 = joint forces due to X_3, all other displacements prevented.
row 1 = forces at the joint 1 due to the displacement of each joint.
row 2 = forces at the joint 2 due to the displacement of each joint.
row 3 = forces at the joint 3 due to the displacement of each joint.

$$K = \begin{bmatrix} (AE/L)_1 & -(AE/L)_1 & 0 \\ -(AE/L)_1 & (AE/L)_1 + (AE/L)_2 & -(AE/L)_2 \\ 0 & -(AE/L)_2 & (AE/L)_2 \end{bmatrix}$$

Note that the off diagonal terms represent the member stiffnesses (or the stiffness between joints) and the terms on the main diagonal represent the joint stiffnesses (which are equal to the sum of the stiffnesses of the members attached to that joint).

$$K = E/L \begin{bmatrix} A_1 & -A_1 & \\ -A_1 & A_1 + A_2 & -A_2 \\ & -A_2 & A_2 \end{bmatrix}$$

$$P = KX, \quad \begin{Bmatrix} R_1 \\ 0 \\ 1000 \end{Bmatrix} = E/5 \begin{bmatrix} 4 & 4 & 0 \\ -4 & 6 & -2 \\ 0 & -2 & 2 \end{bmatrix} \begin{Bmatrix} X_1 \\ X_2 \\ X_3 \end{Bmatrix}$$

Applying the boundary conditions:

$$\begin{Bmatrix} R_1 \\ 0 \\ 1000 \end{Bmatrix} = E/5 \begin{bmatrix} 4 & 4 & 0 \\ -4 & 6 & -2 \\ 0 & -2 & 2 \end{bmatrix} \begin{Bmatrix} X_1 \\ X_2 \\ X_3 \end{Bmatrix}$$

$$K = E/5 \begin{bmatrix} 6 & -2 \\ -2 & 2 \end{bmatrix}, \quad D = 8$$

$$K^{-1} = \frac{5}{8E} \begin{bmatrix} 2 & 2 \\ 2 & 6 \end{bmatrix}$$

$$\frac{5}{8E} \begin{bmatrix} 2 & 2 \\ 2 & 6 \end{bmatrix} \begin{Bmatrix} 0 \\ 1000 \end{Bmatrix} = \begin{Bmatrix} X_2 \\ X_3 \end{Bmatrix}$$

$$\frac{5}{8E}(2)(1000) = X_2 = 1250/E$$

$$\frac{5}{8E}(6)(1000) = X_3 = 3750/E$$

$$X_3 - X_2 = 2500/E$$

$$S_{23} = (X_3 - X_2)E/L = 2500/E(E/5) = \underline{\textbf{500 psi}}$$

$$S_{12} = X_2 E/L = 1250E/(5E) = \underline{\textbf{250 psi}}$$

Example Problem 2-2. Using the matrix method, determine the stresses in the three columns of the structure shown in Figure (2-9). Find the reactions and perform an equilibrium check. The material is concrete with a modulus of elasticity of 3,000 ksi. $A_1 = A_2 = A_3 = 20$ sq. in.

Joints (1) and (2) may translate only in the Y direction, joints (3) and (4) are fixed.

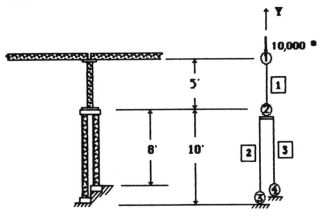

Real Structure   Idealized Structure

FIG. 2-9

The stiffness matrix is:

$$K = \begin{bmatrix} \left(\dfrac{AE}{L}\right)_1 & -\left(\dfrac{AE}{L}\right)_1 & 0 & 0 \\[2mm] -\left(\dfrac{AE}{L}\right)_1 & \left(\dfrac{AE}{L}\right)_1 + \left(\dfrac{AE}{L}\right)_2 + \left(\dfrac{AE}{L}\right)_3 & -\left(\dfrac{AE}{L}\right)_2 & -\left(\dfrac{AE}{L}\right)_3 \\[2mm] 0 & -\left(\dfrac{AE}{L}\right)_2 & \left(\dfrac{AE}{L}\right)_2 & 0 \\[2mm] 0 & -\left(\dfrac{AE}{L}\right)_3 & 0 & \left(\dfrac{AE}{L}\right)_3 \end{bmatrix}$$

The stiffness matrix, after applying the boundary conditions, is:

$$K = \begin{bmatrix} -\left(\dfrac{AE}{L}\right)_1 & -\left(\dfrac{AE}{L}\right)_1 \\[2mm] -\left(\dfrac{AE}{L}\right)_1 & \left(\dfrac{AE}{L}\right)_1 + \left(\dfrac{AE}{L}\right)_2 + \left(\dfrac{AE}{L}\right)_3 \end{bmatrix}$$

$$\left(\frac{AE}{L}\right)_1 = \frac{20 \times 3000}{60} = 1000 \text{k/in}$$

$$\left(\frac{AE}{L}\right)_2 = \frac{20 \times 3000}{96} = 625 \qquad K = \begin{bmatrix} 1000 & -1000 \\ -1000 & 1000 \end{bmatrix}$$

$$\left(\frac{AE}{L}\right)_3 = \frac{20 \times 3000}{120} = 500$$

$$D = 1000\,(2125) - (-1000)(-1000) = 1,125,000$$

$$K^{-1} = \frac{1}{1,125,000} \begin{bmatrix} 2125 & 1000 \\ 1000 & 1000 \end{bmatrix}$$

$$\frac{1}{1,125,000}\begin{bmatrix} 2125 & 1000 \\ 1000 & 1000 \end{bmatrix}\begin{Bmatrix} 10 \\ 0 \end{Bmatrix} = \begin{Bmatrix} Y_1 \\ Y_2 \end{Bmatrix}$$

$$Y_1 = \frac{2125(10)}{1,125,000} = 0.01889''$$

$$Y_2 = \frac{1000(10)}{1,125,000} = 0.00889''$$

$e_1 = (Y_1\text{-}Y_2)/L_1 = (0.01889 - 0.00889)/60 = 0.0001667$

$S_1 = e_1E = (0.001667)(3000) = 0.5001$ ksi , say *500 psi*

$S_2 = e_2E = (0.00889)(3000)/120 = 0.22225$ ksi , say *220psi*

$S_3 = e_3E = (0.00889)(3000)/96 = 0.2778$ ksi , say*280 psi*

$F_2 = S_2A_2 = 0.22225(20) = 4450\# = R_3$, say 4.5 K

$F_3 = S_3A_3 = 0.2778(20) = 5550\# = R_4$ say 5.5 K

Check equilibrium : $R_3 + R_4 = 10$ K

Example Problem 2-3: Set up the equation $P = KX$ in matrix form for the entire structure. Apply the boundary conditions. Find the joint displacements and member forces. Assume positive X is downward.

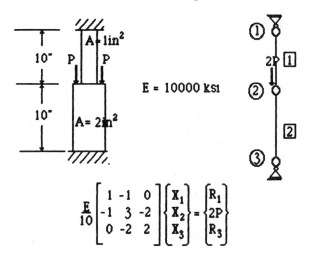

$$\frac{E}{10}\begin{bmatrix} 1 & -1 & 0 \\ -1 & 3 & -2 \\ 0 & -2 & 2 \end{bmatrix}\begin{Bmatrix} X_1 \\ X_2 \\ X_3 \end{Bmatrix} = \begin{Bmatrix} R_1 \\ 2P \\ R_3 \end{Bmatrix}$$

Since joints 1 and 3 are restrained, the boundary conditions are that the displacements of joints 1 and 3 are zero. Therefore, delete the first and third rows and columns of the stiffness matrix.

$$\begin{Bmatrix} R_1 \\ 2P \\ R_3 \end{Bmatrix} = \frac{E}{10} \begin{bmatrix} 1 & -1 & \\ -1 & 3 & -2 \\ & -2 & 2 \end{bmatrix} \begin{Bmatrix} X_1 \\ X_2 \\ X_3 \end{Bmatrix}$$

$$\frac{E}{10}\begin{bmatrix}3\end{bmatrix}\{X_2\} = \{2P\}$$

$$\frac{3EX_2}{10} \quad 2P$$

$$X_2 = \frac{20P}{3E} = \frac{20(100)}{30000} = \frac{1}{15} \text{ in}$$

Force in member $1 = R_1$:

$$R_1 = K_1 X_2 = (AE/L)(20P/3E) = 2P/3 = 66.7k$$

Force in member $2 = R_3$:

$$R_3 = K_2 X_2 = (AE/L)(20P/3E) = 4P/3 = 133k$$

Check: $4P/3 + 2P/3 = 2P$, ok

Example Problem 2-4: Using the Matrix Method, determine the value of the member forces. Assume positive X is upward.

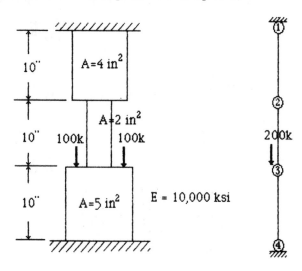

$$\begin{Bmatrix} R_1 \\ 0 \\ -200 \\ R_4 \end{Bmatrix} = \frac{E}{10} \begin{bmatrix} 4 & -4 & 0 & 0 \\ -4 & 6 & -2 & 0 \\ 0 & -2 & 7 & -5 \\ 0 & 0 & -5 & 5 \end{bmatrix} \begin{Bmatrix} X_1 \\ X_2 \\ X_3 \\ X_4 \end{Bmatrix}$$

$$\begin{Bmatrix} 0 \\ -200 \end{Bmatrix} = 1000 \begin{bmatrix} 6 & -2 \\ -2 & 7 \end{bmatrix} \begin{Bmatrix} X_2 \\ X_3 \end{Bmatrix}$$

$$|K| = (6000)(7000) - (-2000)(-2000) = 38 \times 10^6$$

$$10^{-3} \begin{bmatrix} 0.184 & 0.053 \\ 0.053 & 0.158 \end{bmatrix} \begin{Bmatrix} 0 \\ -200 \end{Bmatrix} = \begin{Bmatrix} X_2 \\ X_3 \end{Bmatrix}$$

Solve for joint displacements:

$$184(0) + 53(-200) = X_2 \quad X_2 = -0.0106''$$
$$53(0) + 158(-200) = X_3 \quad X_3 = -0.0316''$$

Solve for member forces and reactions:

$$R_1 = F_{1-2} = [4(10000)/10](0.0106) = 42 \text{ k}$$
$$F_{2-3} = [2(10000)/10](0.0316 - 0.0106) = 42 \text{ k}$$
$$R_4 = F_{3-4} = [5(10000)/10](-0.0316) = 158 \text{ k}$$

Check Equilibrium:

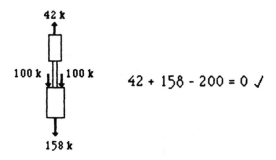

$$42 + 158 - 200 = 0 \checkmark$$

Example Problem 2-5: Determine the fixed-end forces for the structure shown.

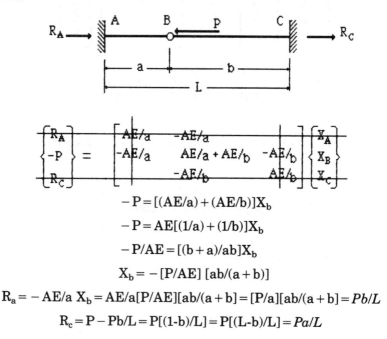

$$-P = [(AE/a) + (AE/b)]X_b$$

$$-P = AE[(1/a) + (1/b)]X_b$$

$$-P/AE = [(b+a)/ab]X_b$$

$$X_b = -[P/AE][ab/(a+b)]$$

$$R_a = -AE/a\ X_b = AE/a[P/AE][ab/(a+b)] = [P/a][ab/(a+b)] = Pb/L$$

$$R_c = P - Pb/L = P[(1-b)/L] = P[(L-b)/L] = Pa/L$$

PROBLEMS FOR SOLUTION:

Solve the following problems using matrix methods

2.1. Determine the stresses in the impact bar shown in Figure (2-10) and perform an equilibrium check.

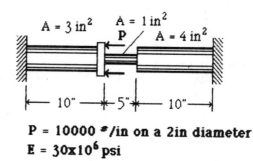

$P = 10000$ #/in on a 2in diameter

$E = 30 \times 10^6$ psi

FIG. 2-10

2.2. Determine the stresses in the three sections of the stepped bar shown.

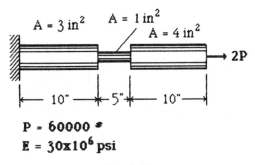

P = 60000 ⁼
E = 30x10⁶ psi

Fig. 2-11

2.3 Find the joint displacements and the forces in the columns of the structure of Figure (2-12). Draw the idealized structure. Consider the truss to be relatively rigid. Discuss how the boundary conditions on the truss (the truss may or may not rotate) will affect the forces in the columns. Es = 30,000 ksi, Ec = 3000 ksi, As = 12 sq. in., Ac = 200 sq. in.

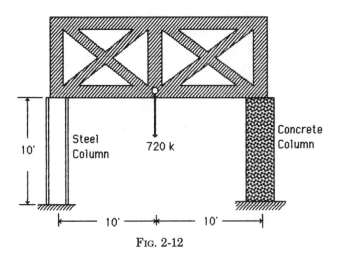

Fig. 2-12

2.4 The steel bolt shown has had its shank reduced in area to better absorb shock loading. The nominal diameter of the bolt is ⅝ in. At two locations, as shown in Figure (1-13), the diameter has been machined to a ½ in. diameter for a distance of 1 in. The two plates are each 1⅝ in.

thick and are being pulled apart with a total force of 100 Kips. Assuming the plates are relatively rigid as compared to the bolt, calculate the stretch in the bolt. Assuming the nut is initially hand tight. $Es = 30,000$ ks

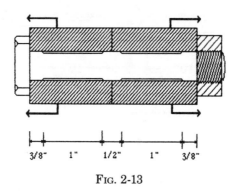

FIG. 2-13

2.5 Determine the displacement and the forces in the concrete and in the steel for the column shown in figure (2-14). $Ec = 3000$ ksi, $E.s = 30000$ ksi

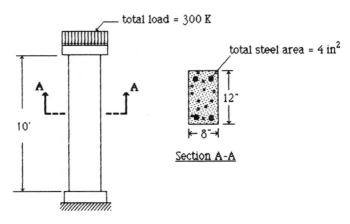

FIG. 2-14

2.6 For the 2 ton crane shown in figure (2-15), find the forces in the three links. The outer links are steel with $E = 30,000$ ksi; the inner link is aluminum with $E = 10,000$ ksi. All links are 10 sq. in. in area. The crane girder may be considered rigid.

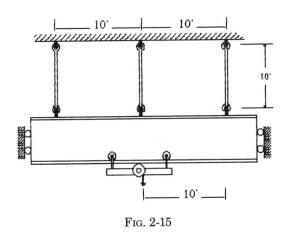

FIG. 2-15

2-7. Using matrix methods, determine the displacement of the hook and the force in each cable. E = 25000 ksi

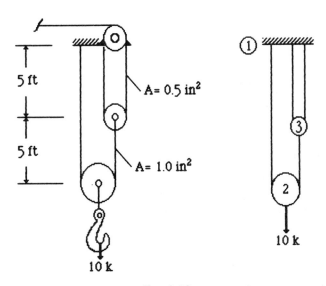

FIG. 2-16

2-8. A steel rod is loaded in tension with one kip at A as shown. The rod is attached to a rigid cap at B, which is supported by an aluminum pipe. The pipe is fixed at C. Determine the value of the displacement of point A. The properties of the rod are: $E = 30,000$ ksi & $A = 2$ sq. in. The properties of the pipe are: $E = 10,000$ ksi, O.D. $= 4''$, & I.D. $= 3.5''$.

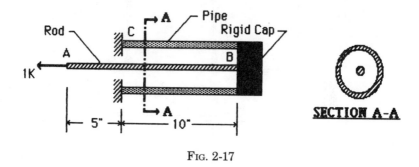

FIG. 2-17

CHAPTER 3

THE SLOPE-DEFLECTION METHOD REVIEWED, SOLUTIONS WITH MATRIX ALGEBRA

The Slope-Deflection Method considers only the moments which result at the joints due to the relative end displacements, rotations, and any external member loads. These moments are calculated separately and summed at each end of each member. The moment equilibrium equations are then written at each joint. The solution is obtained by solving the set of simultaneous equations. Matrix algebra may be used in solving any set of simultaneous equations and will be the method used herein.

The structure shown in Figure (3-1) displaces due to the external loads acting on it. Each member must deform such that its ends translate and rotate consistent with the joints to which it is attached.

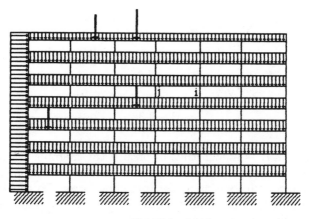

FIG. 3-1

Figure (3-2) shows the displaced shape of joints (i) and (j) and the (i) and (j) ends of all the members attached to these joints. Since the members are rigidly attached to each other at the joints, the angle between any two of them remains unchanged during joint rotation.

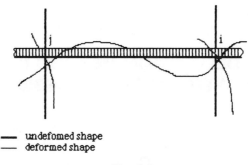

— undefomed shape
— deformed shape

FIG. 3-2

The moments at the ends of member (ij) are the combined result of the rotation of joint (i), the rotation of joint (j), the relative displacement of ends (i) and (j), and the fixed end moments due to the external loads on member (ij),

The following sign convention is adopted for the forces and displacements in the member (local) coordinate system, all positive as shown in Figure (3-3).

FIG. 3-3

In Chapter 2 it is shown that to obtain the member stiffnesses a displacement is imposed at one end of a member, while holding the other end fixed, and calculating the resulting forces. This was done for axial stiffness only. The same procedure will now be used to calculate the flexural stiffness of a beam element.

Remove the external load.

Determine the end moments due to a forced rotation of end (j) while end (i) is held fixed.

Unlocking joint (j) and imposing on it a rotation θ_j:

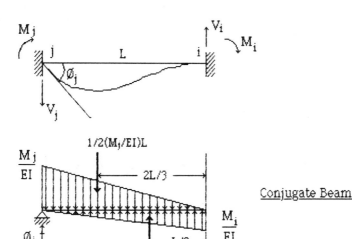

FIG. 3-4

From the conjugate beam:

$\Sigma M_j = 0$: $[\frac{1}{2}(M_j/EI)L]\,(L/3) - [\frac{1}{2}(M_i/EI)]\,(2L/3) = 0$

$$M_i = M_j/2$$

$\Sigma M_i = 0$: $\theta_j L - [\frac{1}{2}(M_j/EI)L]\,(2L/3) + [\frac{1}{2}(M_i/EI)]\,(L/3) = 0$

$$\theta_j = 3M_j L/12EI$$

$$M_j = [4EI/L]\,\theta_j$$

$$M_i = [2EI/L]\,\theta_j$$

From the original beam:

$\Sigma M_j = 0$: $M_j - V_i L + M_i = 0$

$$V_i = [6EI/L^2]\,\theta_j$$

$\Sigma F_Y = 0$: $V_j = [6EI/L^2]\,\theta_j$

Determine the end moments due to a forced rotation of end (i) while end (j) is held fixed:

Unlocking joint (i) and imposing a rotation θ_i will result in:

$$M_i = [4EI/L]\,\theta_i$$

$$M_j = [2EI/L]\,\theta_i$$

With both joints locked against rotation, impose a relative transverse displacement between joints (i) and (j). Imposing a displacement Y to joint (i) relative to (j) results in the following forces and moments:

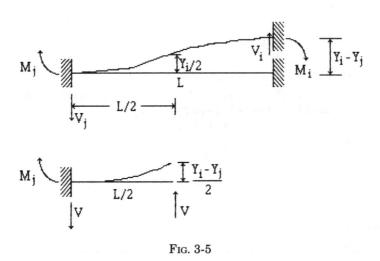

FIG. 3-5

From equilibrium considerations, the value of the shear force in the beam, at any point must be equal to V_i. The half-beam is a simple cantilever with an end load equal to $V_i = V_j = V$. By symmetry, the moment at the center of the beam must equal zero and $M_i = M_j = M = VL/2$.

Then:

$$\frac{(Y_i - Y_j)}{2} = \frac{V(L/2)^3}{3EI}, \quad V = \frac{12EI}{L^3}(Y_i - Y_j)$$

$$M = VL/2, \quad M = \frac{6EI}{L^2}(Y_i - Y_j)$$

The results are the same if end (j) is given a transverse displacement relative to joint (i).

Determine the fixed end moments due to the external loads: (Fixed end moments are listed in Appendix D for some member loads.)

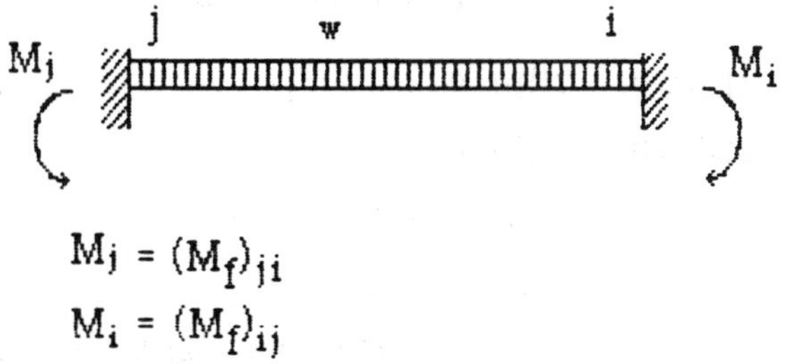

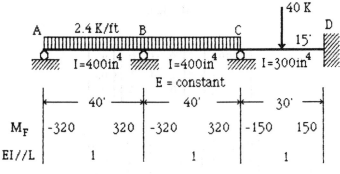

$$M_j = (M_f)_{ji}$$
$$M_i = (M_f)_{ij}$$

FIG. 3-6

The total moments at ends (i) and (j) of member (ij), due to the rotation of end (i); rotation of end (j); relative lateral displacement of ends (i) and (j); and the fixed end moments, have been calculated as:

$$M_i = \left[\frac{4EI}{L}\right](\theta_i) + \left[\frac{2EI}{L}\right](\theta_j) + \left[\frac{6EI}{L^2}\right]\left[(Y_i) - (Y_j)\right] \pm (M_f)_{ij}$$

$$M_j = \left[\frac{2EI}{L}\right](\theta_i) + \left[\frac{4EI}{L}\right](\theta_j) + \left[\frac{6EI}{L^2}\right]\left[(Y_i) - (Y_j)\right] \pm (M_f)_{ij}$$

These moments are calculated for each member of the structure. The the joint equilibrium equations are written and the set of equations solved simultaneously.

Example Problem 3-1 The analysis of the continuous beam of Figure (3-7), by the slope deflection method, results in a set of equations arrived at by writing the moment equations at each side of each joint :

FIG. 3-7

where: M_F = fixed ends moments in K-ft
EI/L = relative stiffness

$$M_{AB} = 4EI/L\ \theta_A + 2EI/L\ \theta_B + \qquad\qquad (M_F)_{AB}$$

$$M_{BA} = 2EI/L\ \theta_A + 4EI/L\ \theta_B + \qquad\qquad (M_F)_{BA}$$

$$M_{BC} = \qquad\quad 4EI/L\ \theta_B + 2EI/L\ \theta_C + \qquad (M_F)_{BC}$$

$$M_{CB} = \qquad\quad 2EI/L\ \theta_B + 4EI/L\ \theta_C + \qquad (M_F)_{CB}$$

$$M_{CD} = \qquad\qquad\qquad 4EI/L\ \theta_C + 2EI/L\ \theta_D + (M_F)_{CD}$$

$$M_{DC} = \qquad\qquad\qquad 2EI/L\ \theta_C + 4EI/L\ \theta_D + (M_F)_{DC}$$

Using the relative stiffnesses, these equations reduce to : ($\theta_D = 0$)

$$M_{AB} = 2\theta_A + \theta_B - 320$$
$$M_{BA} = \ \theta_A + 2\theta_B + 320$$
$$M_{BC} = 2\theta_B + \theta_C - 320$$
$$M_{CB} = \ \theta_B + 2\theta_C + 320$$
$$M_{CD} = 2\theta_C - 150$$
$$M_{DC} = \ \theta_C + 150$$

Writing the moment equilibrium equations at joints A, B, and C, i.e., the sum of the moments at each joint must equal zero, results in three equations with three unknowns.

Sum $M_A = 0$: $M_{AB} = 0$ $2\theta_A + \ \theta_B \qquad\qquad = 320$

Sum $M_B = 0$: $M_{BA} + M_{BC} = 0$ $\theta_A + 4\theta_B + \qquad \theta_C = 0$

Sum $M_C = 0$: $M_{CB} + M_{CD} = 0$ $\theta_B + \qquad 4\theta_C = -170$

These equations may be solved simultaneously using matrix algebra. The necesssary matrix algebra is reviewed in Appendix A.
Putting the equations in matrix form.

$$M = K\ \theta$$

$$\begin{Bmatrix} 320 \\ 0 \\ -170 \end{Bmatrix} = \begin{bmatrix} 2 & 1 & 0 \\ 1 & 4 & 1 \\ 0 & 1 & 4 \end{bmatrix} \begin{Bmatrix} \theta_A \\ \theta_B \\ \theta_C \end{Bmatrix}$$

$$\begin{Bmatrix} \theta_A \\ \theta_B \\ \theta_C \end{Bmatrix} = \begin{bmatrix} 2 & 1 & 0 \\ 1 & 4 & 1 \\ 0 & 1 & 4 \end{bmatrix}^{-1} \begin{Bmatrix} 320 \\ 0 \\ -170 \end{Bmatrix}$$

$$K = \begin{bmatrix} 2 & 1 & 0 \\ 1 & 4 & 1 \\ 0 & 1 & 4 \end{bmatrix}$$

The determinate of $K = 26$.

The minors of K are: $M_{11} = \begin{vmatrix} 4 & 1 \\ 1 & 4 \end{vmatrix} = 15$

$M_{12} = \begin{vmatrix} 1 & 1 \\ 0 & 4 \end{vmatrix} = 4$ $M_{13} = \begin{vmatrix} 1 & 4 \\ 0 & 1 \end{vmatrix} = 1$

$M_{21} = \begin{vmatrix} 1 & 0 \\ 1 & 4 \end{vmatrix} = 4$ $M_{22} = \begin{vmatrix} 2 & 0 \\ 0 & 4 \end{vmatrix} = 8$

$M_{23} = \begin{vmatrix} 2 & 1 \\ 0 & 1 \end{vmatrix} = 2$ $M_{31} = \begin{vmatrix} 1 & 0 \\ 4 & 1 \end{vmatrix} = 1$

$M_{32} = \begin{vmatrix} 2 & 0 \\ 1 & 1 \end{vmatrix} = 2$ $M_{33} = \begin{vmatrix} 2 & 1 \\ 1 & 4 \end{vmatrix} = 7$

The Adjoint of **K** :

$$(-1)^{i+j}[C_{ji}] = \begin{bmatrix} 15 & 4 & 1 \\ -4 & 8 & -2 \\ 1 & -2 & 7 \end{bmatrix}$$

The Inverse of **K** :

$$K^{-1} = \frac{1}{26} \begin{bmatrix} 15 & -4 & 1 \\ - & 8 & -2 \\ 1 & -2 & 7 \end{bmatrix}$$

The solution is :

$$\begin{Bmatrix} \theta_A \\ \theta_B \\ \theta_C \end{Bmatrix} = \frac{1}{26} \begin{bmatrix} 15 & -4 & 1 \\ -4 & 8 & -2 \\ 1 & -2 & 7 \end{bmatrix} \begin{Bmatrix} -320 \\ 0 \\ -170 \end{Bmatrix}$$

$$\theta_A = 1/26[15(320) - 170] = 178.5$$

$$\theta_B = 1/26[-4(320) + 2(170)] = -36.2$$

$$\theta_C = 1/26[320 - 7(170)] = -33.4$$

$$M_{AB} = 2(178.5) + (-36.2) - 320 = 0$$

$$M_{BA} = 178.5 + 2(-36.2) + 320 = 426 \text{ K-ft}$$

$$M_{BC} = 2(-36.2) + (-33.4) - 320 = -426 \text{ K-ft}$$

$$M_{CB} = -36.2 + 2(-33.4) + 320 = 217 \text{ K-ft}$$

$$M_{CD} = 2(-33.4) - 150 = -217 \text{ K-ft}$$

$$M_{DC} = -33.4 + 150 = 116.6 \text{ K-ft}$$

The Slope-Deflection method results in only the joint moments. The shears may now be determined by drawing free body diagrams and applying the static equilibrium equations. If axial loads exist, their effects are neglected. On the other hand, the matrix method produces not only the joint moments, but the shear and axial forces as well.

The matrix method of structural analysis generally refers to not only the solution technique, but also to constructing the equations to be solved in matrix form. This involves the assembling of the load and displacement vectors as well as the stiffness matrix, subjects which are presented in subsequent chapters.

EXAMPLE PROBLEM 3-2

Set up the equations for the moments in the beam shown using slope deflection techniques and solve using matrix methods.

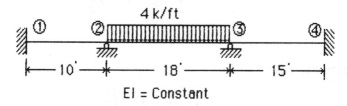

EI = Constant

$$M_{f23} = -4(18)^2/12 = -108 \text{ k-ft}$$
$$M_{f32} = 4(18)^2/12 = 108 \text{ k-ft}$$

$$M_{12} = (2EI/10) \, \theta_2$$
$$M_{21} = (2EI/10) \, 2\theta_2$$
$$M_{23} = (2EI/18) \, (2\theta_2 + \theta_3) - 108$$
$$M_{32} = (2EI/18) \, (2\theta_3 + \theta_2) + 108$$
$$M_{34} = (2EI/15) \, 2\theta_3$$
$$M_{43} = (2EI/15) \, \theta_3$$

$\Sigma M_2 = 0$: $M_{21} + M_{23} = 0$: $(2EI/10) \, 2\theta_2 + (2EI/18) \, (2\theta_2 + \theta_3) - 108 = 0$

$$EI(.311\theta_2 + .0555\theta_3) = 54 \qquad \text{-1-}$$

$\Sigma M_3 = 0$: $M_{32} + M_{34} = 0$: $(2EI/18) \, (2\theta_3 + \theta_2) + 108 + (2EI/15) \, 2\theta_3 = 0$

$$2EI(.0555\theta_2 + .244\theta_3) = -54 \qquad \text{-2-}$$

multiplying equations 1 & 2 by 90:

$$28 \, EI\theta_2 + 5EI\theta_3 = 4860 \qquad \text{-1-}$$
$$5EI\theta_2 + 22EI\theta_3 = -4860 \qquad \text{-2-}$$

$$EI \begin{bmatrix} 28 & 5 \\ 5 & 22 \end{bmatrix} \begin{Bmatrix} \theta_2 \\ \theta_3 \end{Bmatrix} = 4860 \begin{Bmatrix} 1 \\ - \end{Bmatrix}$$

$$|K| = (28)(22) - (5)(5) = 591$$

$$\frac{1}{591EI} \begin{bmatrix} 22 & 5 \\ 5 & 28 \end{bmatrix} \begin{Bmatrix} \theta_2 \\ \theta_3 \end{Bmatrix} = 4860 \begin{Bmatrix} 1 \\ -1 \end{Bmatrix}$$

$$\theta_2 = 4860/591EI(22+5) = 222/EI$$
$$\theta_3 = 4860/591EI(-5-28) = -271/EI$$

$$M_{12} = (2EI/10) \, (222/EI) = 44.4 \text{ k-ft}$$
$$M_{21} = (2EI/10) - 2(222/EI) = 88.8 \text{ k-ft}$$
$$M_{23} = (2EI/18) \, [(2)(222/EI) + (-271/EI)] - 108 = -88.8 \text{ k-ft}$$
$$M_{32} = (2EI/18) \, [(2)(-271/EI) + (222/EI)] + 108 = 72 \text{ k-ft}$$

$$M_{34} = (2EI/15)(\ 2)(-271/EI) = -72 \text{ k-ft}$$

$$M_{43} = (2EI/15)(-271/EI) = -36 \text{ k-ft}$$

EXAMPLE PROBLEM 3-3

Set up the equations for the moments in the frame shown using slope deflection techniques and solve using matrix methods.

Note: For end spans with far end pinned or roller supported:

$$M_{ij} = 3EI/L(\theta_i + \yen) + Mf_i - Mf_j/2$$

where: $\yen = \Delta/L$ & Δ = the relative transverse displacement of end i with respect to end j for member ij. The sign is according to the previously established sign convention.

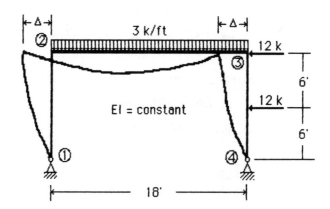

$$Mf_{23} = -wL2/12 = -81 \text{ k-ft}, \quad Mf_{32} = wL2/12 = 81 \text{ k-ft},$$

$$Mf_{34} = -PL/8 = -18 \text{ k-ft}, \quad Mf_{43} = PL/8 = 18 \text{ k-ft}$$

$$M_{21} = EI/4\ (\theta_2 + \yen)$$

$$M_{23} = EI/9\ (2\theta_2 + \theta_3) - 81$$

$$M_{32} = EI/9\ (\theta_2 + 2\theta_3) + 81$$

$$M_{34} = EI/4\ (\theta_3 + \yen) - 18 - 9$$

$$\Sigma M_2 = 0: M_{21} + M_{23} = 0, \ EI/4\ (\theta_2 + \yen) + EI/9\ (2\theta_2 + \theta_3) - 81 = 0 \quad \text{-1-}$$

$$EI(.472\theta_2 + .111\theta_3 + .25\yen) = 81 \quad\quad \text{-1-}$$

$\Sigma M_3 = 0$: $M_{32} + M_{34} = 0$, $EI/9\,(\theta_2 + 2\theta_3) + 81 + EI/4\,(\theta_3 + ¥) - 18 - 9$ -2-

$$EI(.111\theta_2 + .472\theta_3 + .25¥) = -54 \qquad \text{-2-}$$

For an additional equation, equilibrium equations are used on the free body diagrams of the vertical members.

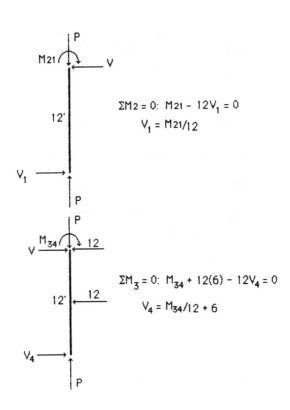

$$\Sigma M_2 = 0: \ M_{21} - 12V_1 = 0$$
$$V_1 = M_{21}/12$$

$$\Sigma M_3 = 0: \ M_{34} + 12(6) - 12V_4 = 0$$
$$V_4 = M_{34}/12 + 6$$

writing the equilibrium equation for the whole structure:

$$\Sigma F_x = 0: \ V_1 + V_4 = 24$$

$$M_{21}/12 + M_{34}/12 + 6 = 24$$

$$M_{21} + M_{34} = 216$$

$$EI/4\,(\theta_2 + ¥) + EI/4\,(\theta_3 + ¥) - 18 - 9 = 216$$

$$EI(.25\theta_2 + .25\theta_3 + .5\yen) = 243 \qquad \text{-3-}$$

$$EI(.472\theta_2 + .111\theta_3 + .25\yen) = 81 \qquad \text{-1-}$$

$$EI(.111\theta_2 + .472\theta_3 + .25\yen) = -54 \qquad \text{-2-}$$

$$EI(.25\theta_2 + .25\theta_3 + .5\yen) = 243 \qquad \text{-3-}$$

$$EI\begin{bmatrix} .472 + .111 + .25 \\ .111 + .472 + .25 \\ .25 + .25 + .5 \end{bmatrix} \begin{Bmatrix} \theta_2 \\ \theta_3 \\ \yen \end{Bmatrix} = \begin{Bmatrix} 81 \\ -54 \\ 243 \end{Bmatrix}$$

$$[K] = (.472)^2(.5) + 2(.111)(.25)^2 - 2(.472)(.25)^2 - (.111)^2(.5)$$

$$[K] = .1114 + .00139 - .059 - .00616 = 0.0601$$

$$M_{11} = (.472)(.5) - (.25)^2 = 0.1735$$

$$M_{12} = (.111)(.5) - (.25)^2 = -0.007$$

$$M_{13} = (.111)(.25) - (.25)(.472) = -0.0903$$

$$M_{21} = M_{12}$$

$$M_{22} = (.472)(.5) - (.25)^2 = 0.1735$$

$$M_{23} = (.472)(.25) - (.25)(.111) = 0.0903$$

$$M_{31} = M_{13}$$

$$M_{32} = M_{23}$$

$$M_{33} = (.472)^2 - (.111)^2 = 0.21$$

$$K^{-1} = \frac{1}{.0601EI} \begin{bmatrix} .1735 & .007 & -.0903 \\ .007 & .1735 & -.0903 \\ -.0903 & -.0903 & .21 \end{bmatrix}$$

$$\frac{1}{.0601EI} \begin{bmatrix} .1735 & .007 & -.0903 \\ .007 & .1735 & -.0903 \\ -.0903 & -.0903 & .21 \end{bmatrix} \begin{Bmatrix} 81 \\ -54 \\ 243 \end{Bmatrix} \begin{Bmatrix} \theta_2 \\ \theta_3 \\ \yen \end{Bmatrix}$$

$$1/0.0601EI\ [(.1735)(81)+(.007)(-54)+(-.0903)(243)]=\theta_2$$

$$\theta_2=-144/EI$$

$$1/0.0601EI\ [(.007)(81)+(.1735)(-54)+(-.0903)(243)]=\theta_3$$

$$\theta_3=-512/EI$$

$$1/0.0601EI\ [(-.0903)(81)+(-.0903)(-54)+(.21)(243)]=\yen$$

$$\yen=809/EI$$

$$M_{21}=EI/4(-144/EI+809/EI)=166\text{ k-ft}$$

avg. = 168 k-ft

$$M_{23}=EI/9(2(-144/EI)+(-512/EI))-81=-170\text{ k-ft}$$

$$M_{32}=EI/9(2(-512/EI)+(-144/EI))+81=-49\text{ k-ft}$$

avg. = 48 k-ft

$$M_{34}=EI/4(-512/EI+809/EI)-27=47$$

PROBLEMS FOR SOLUTION:

3-1. Determine the reactions for the beam shown. In addition to the applied loads shown, the support at B settles 0.1 inches. EI is constant.

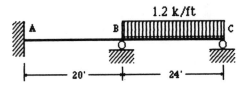

3-2. Determine the reactions for the beam shown.

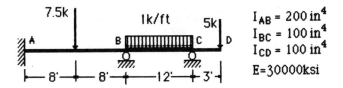

3.3. Determine the moments at the supports. E is constant.

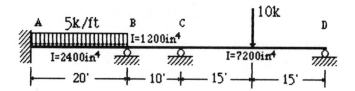

3.4. Determine the joint moments for the portal frame shown. EI is constant.

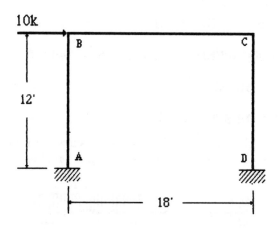

3-5. Determine the moments in the beam shown.

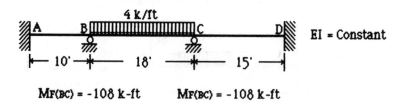

$M_{F(BC)} = -108$ k-ft $M_{F(BC)} = -108$ k-ft

3-6. Calculate the reaction forces on the frame shown. $EI = 120 \times 10^8$ psi

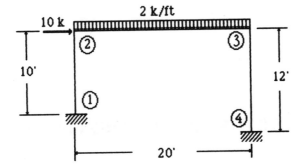

3-7. Determine the moments at the joints. $E = 30000$ ksi

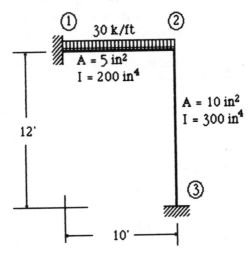

CHAPTER 4

THE STRUCTURAL STIFFNESS MATRIX

In Chapter 2, the stiffness terms of an axial force member were developed. In Chapter 3, the stiffness terms were developed for a beam element with end conditions such that the rotation and translation of the joints transmit to the member shear forces and bending moments. These terms are now combined to form the element stiffness matrix for a **Plane Frame Element.**

$$
\begin{Bmatrix} P_i \\ V_i \\ M_i \\ P_j \\ V_j \\ M_j \end{Bmatrix}
=
\begin{bmatrix}
\dfrac{AE}{L} & 0 & 0 & \dfrac{AE}{L} & 0 & 0 \\[2mm]
0 & \dfrac{12EI}{L^3} & \dfrac{6EI}{L^2} & 0 & \dfrac{12EI}{L^3} & \dfrac{6EI}{L^2} \\[2mm]
0 & \dfrac{6EI}{L^2} & \dfrac{4EI}{L} & 0 & \dfrac{6EI}{L^2} & \dfrac{2EI}{L} \\[2mm]
\dfrac{AE}{L} & 0 & 0 & \dfrac{AE}{L} & 0 & 0 \\[2mm]
0 & \dfrac{12EI}{L^3} & \dfrac{6EI}{L^2} & 0 & \dfrac{12EI}{L^3} & \dfrac{6EI}{L^2} \\[2mm]
0 & \dfrac{6EI}{L^2} & \dfrac{2EI}{L} & 0 & \dfrac{6EI}{L^2} & \dfrac{4EI}{L}
\end{bmatrix}
\begin{Bmatrix} X_i \\ Y_i \\ \theta_i \\ X_j \\ Y_j \\ \theta_j \end{Bmatrix}
$$

The matrix of stiffness terms is called the stiffness matrix of element (ij), or the element stiffness matrix for element (ij). Which may be written:

$$\begin{Bmatrix} P_i \\ V_i \\ M_i \\ P_j \\ V_j \\ M_j \end{Bmatrix} = \begin{bmatrix} k_{ii} & \vline & k_{ij} \\ \hline k_{ji} & \vline & k_{jj} \end{bmatrix} \begin{Bmatrix} X_i \\ Y_i \\ \theta_i \\ X_j \\ Y_j \\ \theta_j \end{Bmatrix}$$

k_{ii} = the forces at (i) due to the displacements at (i) = the stiffness of joint i.

k_{ij} = the forces at (i) due to the displacements at (j) = the stiffness of member ij.

k_{ji} = the forces at (j) due to the displacements at (i) = the stiffness of member ji.

k_{jj} = the forces at (j) due to the displacements at (j) = the stiffness of joint j.

SIGN CONVENTION AND COORDINATE SYSTEMS:

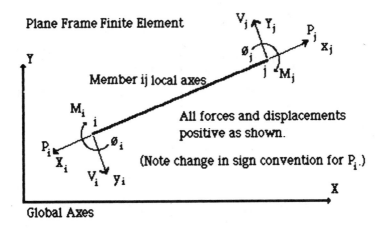

Plane Frame Finite Element

Member ij local axes

All forces and displacements positive as shown.

(Note change in sign convention for P_i.)

Global Axes

Fɪɢ. 4-1

Example Problem 4-1. Form the member stiffness matrices for the structure shown. $E = 30000$ ksi

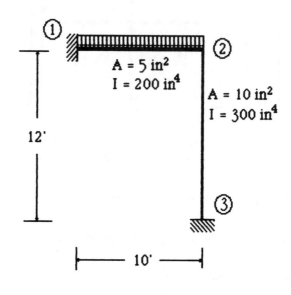

FIG. 4-2

$$k_{21} = E/L \begin{bmatrix} A & 0 & 0 \\ 0 & 12I/L^2 & 6I/L \\ 0 & 6I/L & 2I \end{bmatrix} = \frac{30000}{120} \begin{bmatrix} 5 & 0 & 0 \\ 0 & \dfrac{2400}{120} & \dfrac{1200}{120} \\ 0 & \dfrac{1200}{120} & 400 \end{bmatrix}$$

$$k_{21} = .25 \times 10^3 \begin{bmatrix} 5 & 0 & 0 \\ 0 & .167 & 10 \\ 0 & 10 & 400 \end{bmatrix} = 10^3 \begin{bmatrix} 1.25 & 0 & 0 \\ 0 & .042 & 2.5 \\ 0 & 2.5 & 100 \end{bmatrix}$$

$$k_{23} = E/L \begin{bmatrix} A & 0 & 0 \\ 0 & 12I/L^2 & 6I/L \\ 0 & 6I/L & 2I \end{bmatrix} = \frac{30000}{144} \begin{bmatrix} 10 & 0 & 0 \\ 0 & \dfrac{3600}{144^2} & \dfrac{1800}{144} \\ 0 & \dfrac{1800}{144} & 600 \end{bmatrix}$$

$$k_{23} = .2083 \times 10^3 \begin{bmatrix} 10 & 0 & 0 \\ 0 & .1736 & 12.5 \\ 0 & 12.5 & 600 \end{bmatrix} = 10^3 \begin{bmatrix} 2.083 & 0 & 0 \\ 0 & .036 & 2.6 \\ 0 & 2.6 & 125 \end{bmatrix}$$

Let us examine each 2-2 term:

In k_{21}, the 2-2 term is the shear force (perpendicular to member 21) due to a relative transverse displacement of the ends of member 21, which is in the global Y direction. In k_{23}, the 2-2 term is a shear force developed (perependicular to member 23) due to a relative transverse translation of the ends of member 23, which is in the global X direction. Therefore, they may not be added directly, but must be rotated into a common coordinate system prior to being added.

It was demonstrated in Chapter 2 that the joint stiffness is the sum of the member stiffnesses attached to that joint.

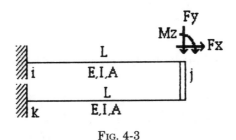

FIG. 4-3

For the structure of Figure (4-3), the stiffness matrix is:

$$
\begin{bmatrix}
k_{ii} & k_{ij} & \\
k_{ij} & k_{jj} & k_{jk} \\
 & k_{jk} & k_{kk}
\end{bmatrix}
$$

where:

$$
k_{ii} = \begin{bmatrix}
\dfrac{AE}{L} & 0 & 0 \\
0 & \left[\dfrac{12EI}{L^3}\right] & \left[\dfrac{6EI}{L^2}\right] \\
0 & \left[\dfrac{6EI}{L^2}\right] & \left[\dfrac{4EI}{L}\right]
\end{bmatrix}_{ij}
, \quad k_{ji} = k_{ij} = \begin{bmatrix}
\dfrac{AE}{L} & 0 & 0 \\
0 & \left[\dfrac{12EI}{L^3}\right] & \left[\dfrac{6EI}{L^2}\right] \\
0 & \left[\dfrac{6EI}{L^2}\right] & \left[\dfrac{2EI}{L}\right]
\end{bmatrix}_{ij}
$$

$$
k_{kk} = \begin{bmatrix}
\dfrac{AE}{L} & 0 & 0 \\
0 & \left[\dfrac{12EI}{L^3}\right] & \left[\dfrac{6EI}{L^2}\right] \\
0 & \left[\dfrac{6EI}{L^2}\right] & \left[\dfrac{4EI}{L}\right]
\end{bmatrix}_{kj}
, \quad k_{jk} = k_{kj} = \begin{bmatrix}
\dfrac{AE}{L} & 0 & 0 \\
0 & \left[\dfrac{12EI}{L^3}\right] & \left[\dfrac{6EI}{L^2}\right] \\
0 & \left[\dfrac{6EI}{L^2}\right] & \left[\dfrac{2EI}{L}\right]
\end{bmatrix}_{jk}
$$

$$
k_{jj} = \begin{bmatrix}
\dfrac{AE}{L} & 0 & 0 \\
0 & \left[\dfrac{12EI}{L^3}\right] & \left[\dfrac{6EI}{L^2}\right] \\
0 & \left[\dfrac{6EI}{L^2}\right] & \left[\dfrac{4EI}{L}\right]
\end{bmatrix}_{ji}
+ \begin{bmatrix}
\dfrac{AE}{L} & 0 & 0 \\
0 & \left[\dfrac{12EI}{L^3}\right] & \left[\dfrac{6EI}{L^2}\right] \\
0 & \left[\dfrac{6EI}{L^2}\right] & \left[\dfrac{2EI}{L}\right]
\end{bmatrix}_{jk}
$$

The mathematical model for the structure of Figure (4-3) is:

$$
\begin{Bmatrix}
F_i \\
F_j \\
F_k
\end{Bmatrix}
=
\begin{bmatrix}
k_{ij} & k_{ij} & \\
k_{ij} & k_{jj} & k_{jk} \\
 & k_{jk} & k_{kk}
\end{bmatrix}
\begin{Bmatrix}
D_i \\
D_j \\
D_k
\end{Bmatrix}
$$

These equations may be rewritten (Leaving off the symbols for matrices and vectors):

$$F = K D$$

$$F_i = k_{ii}D_i + k_{ij}D_j$$

$$F_j = k_{ij}D_i + k_{jj}D_j + k_{jk}D_k$$

$$F_k = \qquad k_{jk}D_j + k_{kk}D_k$$

For a stiffness term on the main diagonal, the two subscripts must be the same, as this term represents the stiffness at the joint. The off-diagonal terms each have two different subscripts. These terms represent the element stiffnesses or the stiffness between joints.

Note that where there is no element between two joints, there is no element stiffness matrix, but, **for every joint, there must be an element stiffness matrix associated with that joint** or there would be no joint stiffness matrix as **the joint stiffness matrix is the sum of the element stiffness matrices of the elements attached to that joint. There must be a joint stiffness matrix for each and every joint in the structure.**

From Maxwell's Reciprocal Theorum, $kij = kji$, for i not equal to j, or the forces at joint i due to the displacements at joint j are equal to the forces at joint j due to the displacements of joint i.

COORDINATE TRANSFORMATION:

In the case of the structure of Figure (4-4), it would seem that the stiffness terms add directly at joint j. This is not the case, however, if the members are not parallel or colinear.

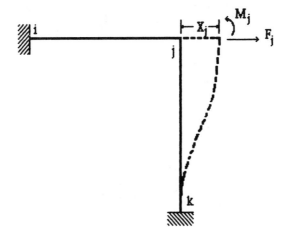

FIG. 4-4

If the two members, ij and jk, are at a right angle to one another, as shown in Figure (4-4), then imposing a displacement to joint j in the X direction would cause an axial force in member ij and a shear force and a moment in member jk. Imposing a displacement to joint j in the Y direction would cause an axial force in member jk and a shear force and a moment in member ij. *The rotational terms would be unaffected.*

If the members are not parallel, the angle between them must be considered and only the parts of the stiffnesses that have common orientation may be added directly. Since stiffnesses are determined using forces and corresponding displacements, *both of which are vector quantities*, the forces and displacements for each member must be brought into a common coordinate system before they can be added directly.

In Figure (4-5), the vectors FX and FY are in the global X-Y system. To rotate them into a local x-y system, their components along the x and y axes are taken and then added directly to obtain the local forces fx and fy in the local x-y system.

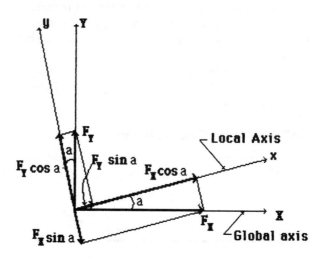

Fig. 4-5

$$f_x = F_X \cos(a) + F_Y \sin(a)$$
$$f_y = -F_X \sin(a) + F_Y \cos(a)$$

$$\left\{ \begin{matrix} f_x \\ f_y \end{matrix} \right\} = \left[\begin{matrix} \cos a & \sin a \\ -\sin a & \cos a \end{matrix} \right] \left\{ \begin{matrix} F_X \\ F_Y \end{matrix} \right\}$$

$$\text{or} \quad \{f\} = [a]\{F\}$$

Where [a] is called the rotation matrix or transformation matrix, {f} is the vector of forces in the local coordinate system, and {F} is the vector of forces in the global coordinate system.

Since forces and displacements are both vector quantities, they may be transformed in the same manner.

$$\begin{Bmatrix} x \\ y \end{Bmatrix} = \begin{bmatrix} \cos a & \sin a \\ -\sin a & \cos a \end{bmatrix} \begin{Bmatrix} X \\ Y \end{Bmatrix}$$

$$\text{or} \quad \{x\} = [a]\{X\}$$

Since the moments and rotations are unaffected by the matrix transformation, the complete transfomation of forces is:

$$\begin{Bmatrix} P \\ V \\ M \end{Bmatrix} = \begin{bmatrix} \cos a & \sin a & 0 \\ -\sin a & \cos a & 0 \\ 0 & 0 & 1 \end{bmatrix} \begin{Bmatrix} F_X \\ F_Y \\ M_Z \end{Bmatrix}$$

Note that the forces and displacements at end i are transformed by aij while those at end j are transformed by aji [see Fig.(4-1)].

$$a_{ij} = \begin{bmatrix} \cos a_{ij} & \sin a_{ij} & 0 \\ -\sin a_{ij} & \cos a_{ij} & 0 \\ 0 & 0 & 1 \end{bmatrix}, \quad a_{ij} = \begin{bmatrix} \cos a_{ji} & \sin a_{ji} & 0 \\ -\sin a_{ji} & \cos a_{ji} & 0 \\ 0 & 0 & 1 \end{bmatrix}$$

$$\cos a_{ij} = (X_i - X_j)/L_{ij} \text{ and } \sin a_{ij} = (Y_i - Y_j)/L_{ij}$$

$$\cos a_{ji} = (X_j - X_i)/L_{ij} \text{ and } \sin a_{ji} = (Y_j - Y_i)/L_{ij}$$

$$\begin{aligned} \cos a_{ji} &= -\cos a_{ij} \\ \sin a_{ji} &= -\sin a_{ij} \end{aligned} \Big\} \; a_{ji} = a_{ij} + 180°$$

The matrix a_{ji} may then be put in terms of the angle a_{ij}:

$$a_{ji} = \begin{bmatrix} -\cos a_{ij} & -\sin a_{ij} & 0 \\ \sin a_{ij} & -\cos a_{ij} & 0 \\ 0 & 0 & 1 \end{bmatrix}$$

Example 4-2 Form the rotation matrix for member 7-8 and member 8-7 of Fig. (4-6).

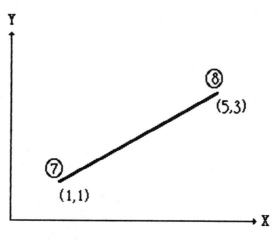

Fig. 4-6

$$L_{7-8} = [(X_7 - X_8)^2 + (Y_7 - Y_8)^2]^{1/2} = [16 + 4]^{1/2} = [20]^{1/2} = 4.472$$

$$\cos a_{7\text{-}8} = (X_7 - X_8)/L_{7\text{-}8} = (1 - 5)/4.472 = -0.894$$

$$\sin a_{7\text{-}8} = (Y_7 - Y_8)/L_{7\text{-}8} = (1 - 3)/4.472 = -0.447$$

$$a_{78} = \begin{bmatrix} \cos a_{78} & -\sin a_{78} & 0 \\ \sin a_{78} & \cos a_{78} & 0 \\ 0 & 0 & 1 \end{bmatrix} = \begin{bmatrix} -.894 & .447 & 0 \\ -.447 & -.894 & 0 \\ 0 & 0 & 1 \end{bmatrix}$$

$$\cos a_{8\text{-}7} = (X_8 - X_7)/L_{7\text{-}8} = (5 - 1)/4.472 = 0.894$$

$$\sin a_{8\text{-}7} = (Y_8 - Y_7)/L_{7\text{-}8} = (3 - 1)/4.472 = 0.447$$

$$a_{87} = \begin{bmatrix} \cos a_{87} & -\sin a_{87} & 0 \\ \sin a_{87} & \cos a_{87} & 0 \\ 0 & 0 & 1 \end{bmatrix} = \begin{bmatrix} .894 & .447 & 0 \\ .447 & -.894 & 0 \\ 0 & 0 & 1 \end{bmatrix}$$

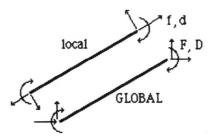

Writing the equations relating the forces and displacements in the global coordinate system to the element stiffness matrix, the global forces and displacements will be rotated into the local coordinate system by premultiplying the force vector and the displacement vector by the rotation matrix. $f = kd$, $f = aF$, $d = aD$

$$f = aF = k a D$$

Where $f = aF =$ forces in the local system, D represents global displacements, and aD represents the displacements in the local system.

$$f_{ij} = a_{ij} F_{ij} = k_{ii} a_{ij} D_i + k_{ij} a_{ji} D_j$$
$$f_{ji} = a_{ji} F_{ji} = k_{ji} a_{ij} D_i + k_{jj} a_{ji} D_j$$

Premultiplying both sides of the equation by the inverse of a:

$$a_{ij}^{-1} a_{ij} F_{ij} = a_{ij}^{-1} k_{ii} a_{ij} D_i + a_{ij}^{-1} k_{ij} a_{ji} D_j$$
$$a_{ji}^{-1} a_{ji} F_{ji} = a_{ji}^{-1} k_{ji} a_{ij} D_i + a_{ji}^{-1} k_{jj} a_{ji} D_j$$

The Global Forces: $\quad F_{ij} = a_{ij}^{-1} k_{ii} a_{ij} D_i + a_{ij}^{-1} k_{ij} a_{ji} D_j$
$$F_{ji} = a_{ji}^{-1} k_{ji} a_{ij} D_i + a_{ji}^{-1} k_{jj} a_{ji} D_j$$

Let: $\quad J_{ij} = a_{ij}^{-1} k_{ii} a_{ij} , \quad S_{ij} = a_{ij}^{-1} k_{ij} a_{ji} ,$ *

$$S_{ji} = a_{ji}^{-1} k_{ji} a_{ij} , \quad J_{ji} = a_{ji}^{-1} k_{jj} a_{ji} ,$$

$$a_{ij}^{-1} a_{ij} = a_{ji}^{-1} a_{ji} = I \text{ (a unit matrix)}$$

then:

$$F_{ij} = J_{ij} D_i + S_{ij} D_j$$
$$F_{ji} = S_{ji} D_i + J_{ji} D_j$$

* Note the order of the subscripts

The S_{ij} matrix may be thought of as the stiffness matrix of member ij rotated into the global coordinate system since it relates global forces to global displacements. Likewise, the J_{ij} matrix may be thought of as the stiffness at joint i, of member ij, in the global coordinate system.

$$J_{ij} = a_{ij}^{-1}\, k_{ii}\, a_{ij}$$

It may be shown that $a_{ij}^{-1} = a_{ij}^{t}$

$$a_{ij}{}^{t} = \begin{bmatrix} \cos a_{ij} & -\sin a_{ij} & 0 \\ \sin a_{ij} & \cos a_{ij} & 0 \\ 0 & 0 & 1 \end{bmatrix}$$

$$a_{ji}{}^{t} = \begin{bmatrix} -\cos a_{ij} & -\sin a_{ij} & 0 \\ \sin a_{ij} & \cos a_{ij} & 0 \\ 0 & 0 & 1 \end{bmatrix}$$

$$J_{ij} = \qquad a_{ij}{}^{5} \qquad\qquad k_{ii} \qquad\qquad a_{ij}$$

$$J_{ij} = \begin{bmatrix} \cos a_{ij} & -\sin a_{ij} & 0 \\ \sin a_{ij} & \cos a_{ij} & 0 \\ 0 & 0 & 1 \end{bmatrix} \begin{bmatrix} AE/L & 0 & 0 \\ 0 & 12EI/L^3 & 6EI/L^2 \\ 0 & 6EI/L^2 & 4EI/L \end{bmatrix} \begin{bmatrix} \cos a_{ij} & \sin a_{ij} & 0 \\ -\sin a_{ij} & \cos a_{ij} & 0 \\ 0 & 0 & 1 \end{bmatrix}$$

$$J_{ij} = \begin{bmatrix} \cos a_{ij} & -\sin a_{ij} & 0 \\ \sin a_{ij} & \cos a_{ij} & 0 \\ 0 & 0 & 1 \end{bmatrix} \begin{bmatrix} (AE/L)\cos a_{ij} & (AE/L)\sin a_{ij} & 0 \\ -(12EI/L^3)\sin a_{ij} & (12EI/L^3)\cos a_{ji} & 6EI/L^2 \\ -(6EI/L^2)\sin a_{ij} & (6EI/L^2)\cos a_{ij} & 4EI/L \end{bmatrix}$$

$$J_{ij} = \begin{bmatrix} (EA/L)\cos^2 a + (12EI/L^3)\sin^2 a & (EA/L - 12EI/L^3)(\cos a \sin a) & -(6EI/L^2)(\sin a) \\ (EA/L - 12EI/L^3)(\cos a \sin a) & (EA/L)(\sin^2 a) + (12EI/L^3)(\cos^2 a) & (6EI/L^2)(\cos a) \\ -(6EI/L^2)(\sin a) & (6EI/L^2)(\cos a) & 4EI/L \end{bmatrix}$$

$$a = a_{ij}$$

$$J_{ji} = a_{ji}^{-1}\, k_{jj}\, a_{ji}$$

$$J_{ji} = \begin{bmatrix} -\cos a_{ij} & \sin a_{ij} & 0 \\ -\sin a_{ij} & -\cos a_{ij} & 0 \\ 0 & 0 & 1 \end{bmatrix} \begin{bmatrix} AE/L & 0 & 0 \\ 0 & 12EI/L^3 & 6EI/L^2 \\ 0 & 6EI/L^2 & 4EI/L \end{bmatrix} \begin{bmatrix} -\cos a_{ij} & -\sin a_{ij} & 0 \\ \sin a_{ij} & -\cos a_{ij} & 0 \\ 0 & 0 & 1 \end{bmatrix}$$

$$J_{ji} = \begin{bmatrix} (EA/L)(\cos^2 a + (12EI/L^3)(\sin^2 a)) & (EA/L - 12EI/L^3)(\cos a \sin a) & (6EI/L^2)(\sin a) \\ (EA/L - 12EI/L^3)(\cos a \sin a) & (EA/L)(\sin^2 a + (12EI/L^3)(\cos^2 a)) & -(6EI/L^2)(\cos a) \\ (6EI/L^2)(\sin a) & -(6EI/L^2)(\cos a) & 4EI/L \end{bmatrix}$$

$a = a_{ij}$

$S_{ij} = a_{ij} - 1 \, k_{ij} \, a_{ji}$

$$S_{ij} = \begin{bmatrix} \cos a_{ij} & -\sin a_{ij} & 0 \\ \sin a_{ij} & \cos a_{ij} & 0 \\ 0 & 0 & 1 \end{bmatrix} \begin{bmatrix} AE/L & 0 & 0 \\ 0 & 12EI/L^3 & 6EI/L^2 \\ 0 & 6EI/L^2 & 2EI/L \end{bmatrix} \begin{bmatrix} -\cos a_{ij} & -\sin a_{ij} & 0 \\ \sin a_{ij} & -\cos a_{ij} & 0 \\ 0 & 0 & 1 \end{bmatrix}$$

$$S_{ij} = \begin{bmatrix} \cos a_{ij} & -\sin a_{ij} & 0 \\ \sin a_{ij} & \cos a_{ij} & 0 \\ 0 & 0 & 1 \end{bmatrix} \begin{bmatrix} -(AE/L)(\cos a_{ij}) & -(AE/L)(\sin a_{ij}) & 0 \\ (12EI/L^3)(\sin a_{ij}) & -(12EI/L^3)(\cos a_{ij}) & 6EI/L^2 \\ (6EI/L^2)(\sin a_{ij}) & -(6EI/L^2)(\cos a_{ij}) & 2EI/L \end{bmatrix}$$

$$S_{ij} = \begin{bmatrix} -[(EA/L)(\cos^2 a) + (12EI/L^3)(\sin^2 a)] & -(EA/L - 12EI/L^3)(\cos a \sin a) & -(6EI/L^2)(\sin a) \\ -(EA/L - 12EI/L^3)(\cos a \sin a) & -[(EA/L)(\sin^2 a + (12EI/L^3)(\cos^2 a)] & (6EI/L^2)(\cos a) \\ (6EI/L^2)(\sin a) & -(6EI/L^2)(\cos a) & 2EI/L \end{bmatrix}$$

$a = a_{ij}$

$S_{ji} = a_{ji}^{-1} \, k_{ji} \, a_{ij}$

$$S_{ij} = \begin{bmatrix} -\cos a_{ij} & \sin a_{ij} & 0 \\ -\sin a_{ij} & -\cos a_{ij} & 0 \\ 0 & 0 & 1 \end{bmatrix} \begin{bmatrix} AE/L & 0 & 0 \\ 0 & 12EI/L^3 & 6EI/L^2 \\ 0 & 6EI/L^2 & 2EI/L \end{bmatrix} \begin{bmatrix} \cos a_{ij} & \sin a_{ij} & 0 \\ -\sin a_{ij} & \cos a_{ij} & 0 \\ 0 & 0 & 1 \end{bmatrix}$$

$$S_{ji} = \begin{bmatrix} -[(EA/L)(\cos^2 a) + (12EI/L^3(\sin^2 a)] & -(EA/L - 12EI/L^3)(\cos a \sin a) & (6EI/L^2)(\sin a) \\ -(EA/L - 12EI/L^3)(\cos a \sin a) & -[(EA/l)(\sin^2 a + (12EI/L^3)(\cos^2 a)] & -(6EI/L^2)(\cos a) \\ -(6EI/L^2)(\sin a) & (6EI/L^2)(\cos a) & 2EI/L \end{bmatrix}$$

$a = a_{ij}$

Note: $S_{ji} = St_{ij}$ and the J_{ij} matrix differs from the S_{ij} matrix only by the signs of the terms in the first two columns and by a factor of two for the 3-3 term. Therefore, the J_{ij} matrix may be obtained by modifying the the S_{ij} matrix by multiplying the first two columns by -1 and the 3-3 term by 2. This will prove very useful later.

It should be noted, at this time, that shear deflections have been ignored in the formulation of the beam stiffness matrix. The result is an element which is stiffer than if the shear terms had been included. However, except for beams which have a depth of cross-section comparable to the beam span, the effect of shear is negligible.

For short deep beams, such as shear walls or gear teeth, the shear terms may be of equal or greater significance than the bending terms.

The displacement of a simple cantilever beam, for instance, is $PL^3/3EI + VL/AG$.

V is the shear force. For a cantilever, $V = P$.

G is the shear modulus of elasticity $= E/2(1 + \mu)$,

A is the cross sectional area.

I is the moment of inertia about the bending axis.

V is the radius of gyration.

μ = poisson's ratio, which for structural metals may be taken as 0.3.

Setting the shear deflection equal to the flexural deflection and solving for the span to depth ratio:

$$PL^3/3EI = VL/AG.$$

$$PL^3/3EI = PL/AE/2.6$$

$$L^2/3I = 2.6/A$$

$$I = Ar^2 , \ L^2/3r^2 = 2.6$$

$$L^2/r^2 = 7.8 , \ (L/r)^2 = 7.8$$

A slenderness ratio of around 7.8 would be unusual for a structural member but not for a machine part.

It may be concluded that the effect of shear on the stiffness of a beam element is not to be ignored, but is rarely significant.

Example Problem 4-3 Form the structural stiffness matrix for the structure shown in Fig. (4-7) and apply the boundary conditions. (see Example Problem 4-1) E = 30,000 ksi

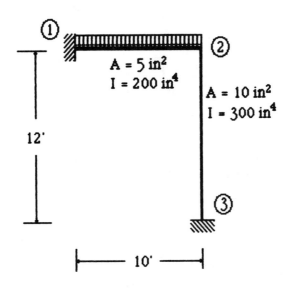

FIG. 4-7

Stiffness Matrix Pattern:

J_1	S_{12}	
S_{21}	J_2	S_{23}
	S_{32}	J_3

$\cos a_{12} = {}^{-1}$, $\sin a_{12} = 0$

$$S_{12} = 10^3 \begin{bmatrix} -1.25 & 0 & 0 \\ 0 & -0.042 & -2.5 \\ 0 & 2.5 & 100.0 \end{bmatrix}$$

$$S_{21} = 10^3 \begin{bmatrix} -1.25 & 0 & 0 \\ 0 & -0.042 & 2.5 \\ 0 & -2.5 & 100.0 \end{bmatrix}$$

$$J_{12} = J_1 = 10^3 \begin{bmatrix} 1.25 & 0 & 0 \\ 0 & 0.042 & -2.5 \\ 0 & -2.5 & 200.0 \end{bmatrix}$$

$\cos a_{23} = 0, \sin a_{23} = 1$

$$S_{23} = 10^3 \begin{bmatrix} -0.036 & 0 & -2.6 \\ 0 & -2.083 & 0 \\ 2.6 & 0 & 125.0 \end{bmatrix}$$

$$S_{32} = 10^3 \begin{bmatrix} -0.036 & 0 & 2.6 \\ 0 & -2.083 & 0 \\ -2.6 & 0 & 125.0 \end{bmatrix}$$

$$J_{32} = J_3 = 10^3 \begin{bmatrix} 0.036 & 0 & 2.6 \\ 0 & 2.083 & 0 \\ 2.6 & 0 & 250.0 \end{bmatrix}$$

$$J_{21} + J_{23} = J_2 = 10^3 \begin{bmatrix} 1.286 & 0 & -2.6 \\ 0 & 2.13 & 2.5 \\ -2.6 & 2.5 & 450.0 \end{bmatrix}$$

$$K = 10^3 \begin{bmatrix} 1.25 & 0 & 0 & -1.25 & 0 & 0 & & & \\ 0 & 0.042 & -2.5 & 0 & -0.042 & 2.5 & & & \\ 0 & -2.5 & 200.0 & 0 & -2.5 & 100.0 & & & \\ -1.25 & 0 & 0 & 1.286 & 0 & -2.6 & -0.036 & -0.036 & -2.6 \\ 0 & -0.042 & -2.5 & 0 & 2.13 & 2.5 & 0 & 0 & 0 \\ 0 & 2.5 & 100.0 & -2.6 & 2.5 & 450.0 & 2.6 & 2.6 & 125.0 \\ & & & -0.036 & 0 & 2.6 & 0.036 & 0.036 & 2.6 \\ & & & 0 & -2.083 & 0 & 0 & 0 & 0 \\ & & & -2.6 & 0 & 125.0 & 2.6 & 2.6 & 250.0 \end{bmatrix}$$

Applying the B.C.:

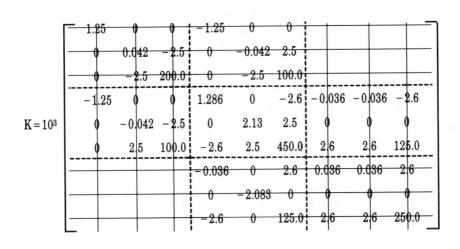

$$K = 10^3 \begin{bmatrix} 1.286 & 0 & -2.6 \\ 0 & 2.13 & 2.5 \\ -2.6 & 2.5 & 450.0 \end{bmatrix}$$

PROCEDURE FOR FORMING
THE STRUCTURAL STIFFNESS MATRIX

1. Draw the idealized structure, numbering all joints and members.

2. Establish the Global Axes.

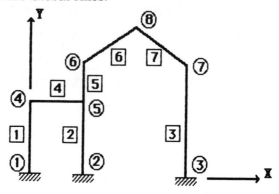

3. Set up the pattern of the stiffness matrix.

$$
K = \begin{array}{|c|c|c|c|c|c|c|c|}
\hline
J_1 & & & S_{14} & & & & \\
\hline
 & J_2 & & & S_{25} & & & \\
\hline
 & & J_3 & & & & S_{37} & \\
\hline
S_{41} & & & J_4 & S_{45} & & & \\
\hline
 & S_{52} & & S_{54} & J_5 & S_{56} & & \\
\hline
\multicolumn{3}{|c}{\text{Symmetric}} & & S_{65} & J_6 & & S_{68} \\
\hline
 & & S_{73} & & & & J_7 & S_{78} \\
\hline
 & & & & & S_{86} & S_{87} & J_8 \\
\hline
\end{array}
$$

4. Calculate each member stiffness, S_{ij}, for $i<j$, and locate the result in its proper place in the stiffness matrix.

$$S_{ij} = a^t_{ij}\, k_{ij}\, a_{ji}$$

5. Determine each S_{ij}, for $i>j$, by taking the corresponding transpose and locate the result in its proper place in the stiffness matrix below the diagonal.

$$S_{ji} = S^t_{ij}, \quad i<j$$

6. Sum all the S_{ij} sub-matrices in each row and place the sum on the main diagonal, modifying the sum by:

    (A) Multiply by (-1) the first and second *columns*.

    (B) Multiply by (2) the last element in the third row.

The result is a joint stiffness matrix.

Note that the joint stiffness matrix J_i equals the sum of the modified member stiffness matrices for members attached to that joint, ie, sum across the rows and modify the result.

$$J_i = \Sigma\, J_{in} = [\,\Sigma a^t_{ij}\, k_{ii}\, a_{ij}\,] = [\Sigma a^t_{ij}\, k_{ij}\, a_{ji}\,]\ (\text{modified}) = \Sigma\, S_{ij}\ (\text{modified})$$

The stiffness matrix has now been formed.

Definitions:

$$k_{ij} = k_{ji} = \begin{bmatrix} AE/L & 0 & 0 \\ 0 & 12EI/L^3 & 6EI/L^2 \\ 0 & 6EI/L^2 & 2EI/L \end{bmatrix}$$

$$a_{ij} = \begin{bmatrix} \cos a & \sin a & 0 \\ -\sin a & \cos a & 0 \\ 0 & 0 & 1 \end{bmatrix}, \quad a^5_{ij} = \begin{bmatrix} \cos a & -\sin a & 0 \\ \sin a & \cos a & 0 \\ 0 & 0 & 1 \end{bmatrix}$$

$$\cos a = \frac{X_i - X_j}{L_{ij}}, \quad \sin a = \frac{Y_i - Y_j}{L_{ij}}$$

(X and Y in the global coordinate system)

$$S_{ij} = a^t_{ij}\, k_{ij}\, a_{ji}$$

$$S_{ij} = \begin{bmatrix} \cos a & -\sin a & 0 \\ \sin a & \cos a & 0 \\ 0 & 0 & 1 \end{bmatrix} \begin{bmatrix} AE/L & 0 & 0 \\ 0 & 12EI/L^3 & 6EI/L^2 \\ 0 & 6EI/L^2 & 2EI/L \end{bmatrix} \begin{bmatrix} -\cos a & -\sin a & 0 \\ \sin a & -\cos a & 0 \\ 0 & 0 & 1 \end{bmatrix}$$

$$S_{ij} = \begin{bmatrix} -\dfrac{EA}{L}\cos^2 a - \dfrac{12EI}{L^3}\sin^2 a & -\dfrac{EA}{L}\sin a \cos a + \dfrac{12EI}{L^3}\cos a \sin a & -\dfrac{6EI}{L^2}\sin a \\[2ex] -\dfrac{EA}{L}\sin a \cos a + \dfrac{12EI}{L^3}\cos a \sin a & -\dfrac{EA}{L}\sin^2 a - \dfrac{12EI}{L^3}\cos^2 a & \dfrac{6EI}{L^2}\cos a \\[2ex] \dfrac{6EI}{L^2}\sin a & -\dfrac{6EI}{L^2}\cos a & \dfrac{2EI}{L} \end{bmatrix}$$

$$a = a_{ij}$$

$$S_{ij} = \begin{bmatrix} -\left(\dfrac{EA}{L}\cos^2 a + \dfrac{12EI}{L^3}\sin^2 a\right) & -\left(\dfrac{EA}{L} - \dfrac{12EI}{L^3}\right)\cos a \sin a & -\dfrac{6EI}{L^2}\sin a \\[2ex] -\left(\dfrac{EA}{L} - \dfrac{12EI}{L^3}\right)\cos a \sin a & -\left(\dfrac{EA}{L}\sin^2 a + \dfrac{12EI}{L^3}\cos^2 a\right) & \dfrac{6EI}{L^2}\cos a \\[2ex] \dfrac{6EI}{L^2}\sin a & -\dfrac{6EI}{L^2}\cos a & \dfrac{2EI}{L} \end{bmatrix}$$

$$a = a_{ij}$$

$$J_{ij} = S_{ij} = \text{modified} \begin{bmatrix} \left(\dfrac{EA}{L}\cos^2 a + \dfrac{12EI}{L^3}\sin^2 a\right) & \left(\dfrac{EA}{L} - \dfrac{12EI}{L^3}\right)\cos a \sin a & -\dfrac{6EI}{L^2}\sin a \\[2ex] \left(\dfrac{EA}{L} - \dfrac{12EI}{L^3}\right)\cos a \sin a & \left(\dfrac{EA}{L}\sin^2 a + \dfrac{12EI}{L^3}\cos^2 a\right) & \dfrac{6EI}{L^2}\cos a \\[2ex] -\dfrac{6EI}{L^2}\sin a & \dfrac{6EI}{L^2}\cos a & \dfrac{2EI}{L} \end{bmatrix}$$

$$a = a_{ij}$$

FORMING THE STRUCTURAL STIFFNESS MATRIX FOR A TRUSS:

The truss element stiffness was derived in Chapter 2. The truss element (rod, link, cable, etc.) can only transmit axial loads. It is pinned at both ends and cannot, therefore, be subjected to shears or moments so long as the external loads are applied between the ends. If, in a structure, there is a member which is not pinned at both ends or if any load is applied to a member, then the structure is not a truss and must be analyzed as a frame.

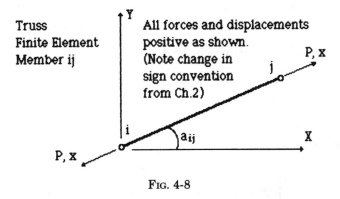

Truss Finite Element Member ij

All forces and displacements positive as shown. (Note change in sign convention from Ch.2)

FIG. 4-8

In order to utilize this element in a structure which has elements which are not parallel, the element stiffness matrices must be rotated into the global coordinate system, in the same manner as for the beam element.

The rotated element stiffness matrix for the beam-column element is:

$$
S_{ij} = \begin{bmatrix}
-[(EA/L)(\cos^2 a) + (12EI/L^3)(\sin^2 a)] & -(EA/L - 12EI/L^3)(\cos a \sin a) & -(6EI/L^2)(\sin a) \\
-(EA/L - 12EI/L^3)(\cos a \sin a) & -[(EA/l)(\sin^2 a + (12EI/L^3)(\cos^2 a)] & (6EI/L^2)(\cos a) \\
(6EI/L^2)(\sin a) & -(6EI/L^2)(\cos a) & 2EI/L
\end{bmatrix}
$$

$a = a_{ij}$

All the bending and shear terms are set to zero for the truss element, giving:

$$
S_{ij} = \begin{bmatrix}
-(EA/L)(\cos^2 a) & -(EA/L)(\cos a \sin a) & 0 \\
-(EA/L)(\cos a \sin a) & -(EA/L)(\sin^2 a) & 0 \\
0 & 0 & 0
\end{bmatrix}
$$

$a = a_{ij}$

In a truss, each joint has only two degrees of freedom, translations in the X and Y directions. There are no joint rotations. Therefore, the third row and column are zeroed out of the matrix, resulting in:

$$S_{ij} = \begin{bmatrix} -(EA/L)(\cos^2 a) & -(EA/L)(\cos a \sin a) \\ -(EA/L)(\cos a \sin a) & -(EA/L)(\sin^2 a) \end{bmatrix}$$

$$a = a_{ij}$$

Which leads to:

$$J_{ij} = \begin{bmatrix} (EA/L)(\cos^2 a) & (EA/L)(\cos a \sin a) \\ (EA/L)(\cos a \sin a) & (EA/L)(\sin^2 a) \end{bmatrix}$$

$$a = a_{ij}$$

Example problem 4-4: For the truss shown in Fig. (4-9), form the structural stiffness matrix and apply the boundary conditions.

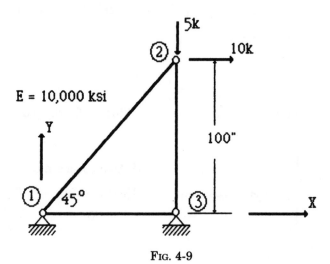

FIG. 4-9

member	A	L	AE/L	cos a	sin a
1–2	1.41	141	100	–.707	–.707
2–3	2.0	100	100	0	1
1–3	2.0	100	200	–1	0

The pattern of the stiffness matrix is:

$$
\begin{Bmatrix} P_x \\ P_y \end{Bmatrix}_1 \begin{Bmatrix} P_x \\ P_y \end{Bmatrix}_2 \begin{Bmatrix} P_x \\ P_y \end{Bmatrix}_3 = \left[\begin{array}{c:c:c} J_1 & S_{12} & S_{13} \\ \hline S_{21} & J_2 & S_{23} \\ \hline S_{31} & S_{32} & J_3 \end{array} \right] \begin{Bmatrix} X \\ Y \end{Bmatrix}_1 \begin{Bmatrix} X \\ Y \end{Bmatrix}_2 \begin{Bmatrix} X \\ Y \end{Bmatrix}_3
$$

Applying the boundary conditions, there are no joint displacements at joints 1 & 3:

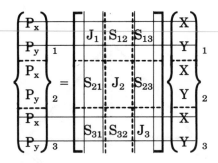

or:

$$
\begin{Bmatrix} P_x \\ P_y \end{Bmatrix}_2 = \begin{bmatrix} J_2 \end{bmatrix} \begin{Bmatrix} X \\ Y \end{Bmatrix}_2
$$

$$
J_2 = (S_{21} + S_{23})_{\text{modified}} = J_{21} + J_{23}
$$

Note that the S matrices are not required in this case.

$$J_{21} = \begin{bmatrix} 100(0.5) & 100(0.5) \\ 100(0.5) & 100(0.5) \end{bmatrix} = 50 \begin{bmatrix} 1 & 1 \\ 1 & 1 \end{bmatrix}$$

$$J_{23} = \begin{bmatrix} 0 & 0 \\ 0 & 200(1.0) \end{bmatrix} = 200 \begin{bmatrix} 0 & 0 \\ 0 & 1 \end{bmatrix}$$

$$J_2 = 50 \begin{bmatrix} 1 & 1 \\ 1 & 5 \end{bmatrix}$$

Example 4-5: For the truss shown in Fig. (4-10), create the structural stiffness matrix and apply the boundary conditions. For all members, AE/L = 1000#/in.

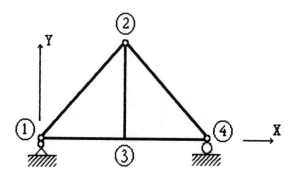

FIG. 4-10

member	cos a	sin a
12	−.707	−.707
31 & 13	−1.0	0
23	0	1.0
24	−.707	.707

$$
K = \begin{array}{|c|c|c|c|}
\hline
J_1 & S_{12} & S_{13} & \\
\hline
S_{21} & J_2 & S_{23} & S_{24} \\
\hline
S_{31} & S_{32} & J_3 & S_{34} \\
\hline
 & S_{42} & S_{43} & J_4 \\
\hline
\end{array}
$$

Applying the boundary conditions. Joint 1 is restrained in both the X and Y directions while joint 4 is restrained only in the Y direction.

$$
K = \begin{array}{|c|c|c|c|}
\hline
J_1 & S_{12} & S_{13} & \\
\hline
S_{21} & J_2 & S_{23} & S_{24} \\
\hline
S_{31} & S_{32} & J_3 & S_{34} \\
\hline
 & S_{42} & S_{43} & J_4 \\
\hline
\end{array}
$$

Note that, in this case, the S matrices are required.

$$S_{12} = 500 \begin{bmatrix} -1 & -1 \\ -1 & -1 \end{bmatrix} = S_{21} \qquad S_{24} = 500 \begin{bmatrix} -1 & 1 \\ 1 & -1 \end{bmatrix} = S_{42}$$

$$S_{13} = 1000 \begin{bmatrix} -1 & 0 \\ 0 & 0 \end{bmatrix} = S_{31} \qquad S_{34} = 1000 \begin{bmatrix} -1 & 0 \\ 0 & 0 \end{bmatrix} = S_{43}$$

$$S_{23} = 1000 \begin{bmatrix} 0 & 0 \\ 0 & -1 \end{bmatrix} = S_{32}$$

$$J_2 = 500 \begin{bmatrix} 1 & 1 \\ 1 & 1 \end{bmatrix} + 1000 \begin{bmatrix} 0 & 0 \\ 0 & 1 \end{bmatrix} + 500 \begin{bmatrix} 1 & -1 \\ -1 & 1 \end{bmatrix} = 1000 \begin{bmatrix} 1 & 0 \\ 0 & 2 \end{bmatrix}$$

$$J_3 = 1000 \begin{bmatrix} 1 & 0 \\ 0 & 0 \end{bmatrix} + 1000 \begin{bmatrix} 0 & 0 \\ 0 & 1 \end{bmatrix} + 1000 \begin{bmatrix} 1 & 0 \\ 0 & 0 \end{bmatrix} = 1000 \begin{bmatrix} 2 & 0 \\ 0 & 1 \end{bmatrix}$$

$$J_4 = 500 \begin{bmatrix} 1 & -1 \\ 1 & -1 \end{bmatrix} + 1000 \begin{bmatrix} 1 & 0 \\ 0 & 0 \end{bmatrix} = 500 \begin{bmatrix} 3 & -1 \\ -1 & 1 \end{bmatrix}$$

$$K = \begin{bmatrix} 1000 & 0 & 0 & 0 & -500 \\ 0 & 2000 & 0 & -1000 & 500 \\ 0 & 0 & 2000 & 0 & -1000 \\ 0 & -1000 & 0 & 1000 & 0 \\ -500 & 500 & -1000 & 0 & 1500 \end{bmatrix}$$

MEMBER RELEASES:

To this point, only structures within which all members are either rigidly connected at the joints (frames) or pinned to the joints (trusses) have been considered. It is not uncommon for structures to have some members rigidly connected to the joints and some members pinned to the joints.

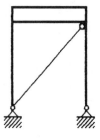 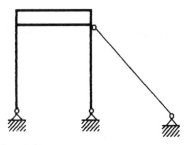

FIG. 4-11

For structures which have all members rigidly connected to each other at the joints (Fig. 4-12), a moment applied to a joint will cause that joint (and possibly others) to rotate. As the joint rotates, each member connected to that joint rotates the same amount and develops a resisting moment and shear. The resisting moment developed in the member is equal to the members rotational stiffness times the joint rotation. The sum of the member moments must equal the moment applied to the joint.

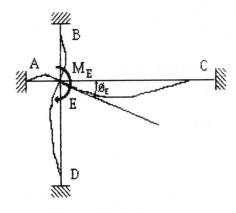

FIG. 4-12

$$M_{EA} = K_{EA}\theta_E$$
$$M_{EB} = K_{EB}\theta_E$$
$$M_{EC} = K_{EB}\theta_E$$
$$M_{EC} = K_{EC}\theta_E$$
$$M_{ED} = K_{ED}\theta_E$$

The sum of the member moments must equal the applied moment.

$$M_{EA} + M_{EB} + M_{EC} + M_{ED} = [K_{EA} + K_{EB} + K_{EC} + K_{ED}]\,\theta_E = M_E = \Sigma K_{Ej}\theta_E$$

$$\text{or,} \qquad M_E/\theta_E = \Sigma K_{Ej} = K_E$$

The stiffness of the joint is equal to the sum of the member stiff-nesses of the members connected to that joint. As derived in previous chapters, the same argument is made for joint displacements and member forces.

If there is a member which is not rigidly connected to a joint to which it is attached, but is pinned to the joint (Fig.4-13), then it does not contribute to the rotational stiffness of the joint. The moment applied to the joint will rotate the joint, but the member which is pinned to that joint will not rotate and therefore will not develop a resisting moment.

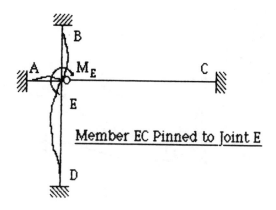

Fɪɢ. 4-13

If the member is attached to the joint with a roller (Fig. 4-15), then motion of the joint in the direction of rolling will not produce any force in the axial direction in the member. In addition to the moment release, the axial force is released. Any degree of freedom may be released at either end of a member, but care must be taken not to create an unstable situation.

THE FIXED-PINNED FINITE ELEMENT:

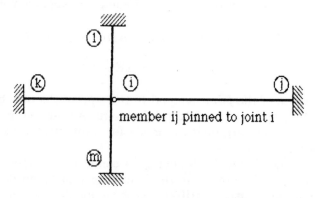

Fɪɢ. 4-14

Imposing an X displacement to each joint separately:

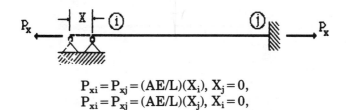

$$P_{xi} = P_{xj} = (AE/L)(X_i), \ X_j = 0,$$
$$P_{xi} = P_{xj} = (AE/L)(X_j), \ X_i = 0,$$

Imposing a Y displacement to each joint i:

$$V_i = V_j = (3EI/L^3)(Y_i) \qquad M_j = V_i\,L = (3EI/L^2)(Y_i)$$

Imposing a Y displacement to joint j:

$$V_i = V_j = (3EI/L^3)(Y_j) \qquad M_j = V_i\,L = (3EI/L^2)(Y_j)$$

Imposing a rotation to joint j:

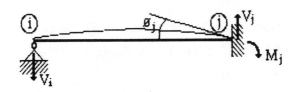

$$V_i = V_j = (3EI/L^2)(\theta_j) \qquad M_j = V_i\,L = (3EI/L)(\theta_j)$$

Imposing a rotation to joint i:

$$V_i = V_j = 0 \qquad\qquad M_j = V_i\,L = 0$$

The moment which rotates joint i does not affect member ij because member ij is pinned to joint i.

The k matrices are:

	X_i	Y_i	θ_i	X_j	Y_j	θ_j
P_i	AE/L	0	0	AE/L	0	0
V_i	0	$3EI/L^3$	0	0	$3EI/L^3$	$3EI/L^2$
M_i	0	0	0	0	0	0
P_j	AE/L	0	0	AE/L	0	0
V_j	0	$3EI/L^3$	0	0	$3EI/L^3$	$3EI/L^2$
M_j	0	$3EI/L^3$	0	0	$3EI/L^3$	$3EI/L$

$$S_{ij} = \begin{bmatrix} -AE/L\cos^2 a - 3EI/L^3\sin^2 a & (-AE/L + 3EI/L^3)\cos a \sin a & -3EI/L^2\sin a \\ (-AE/L + 3EI/L^3)\cos a \sin a & -AE/L\sin^2 a - 3EI/L^3\cos^2 a & 3EI/L^2\cos a \\ 0 & 0 & 0 \end{bmatrix}$$

$$S_{ji} = \begin{bmatrix} -AE/L\cos^2 a - 3EI/L^3\sin^2 a & -(AE/L - 3EI/L^3)\cos a \sin a & 0 \\ -(AE/L - 3EI/L^3)\cos a \sin a & -AE/L\sin^2 a + 3EI/L^3\cos^2 a & 0 \\ -3EI/L^2\sin a & 3EI/L^2\cos a & 0 \end{bmatrix}$$

$$J_{ii} = \begin{bmatrix} AE/L\cos^2 a + 3EI/L^3\sin^2 a & (AE/L - 3EI/L^3)\cos a \sin a & 0 \\ (AE/L - 3EI/L^3)\cos a \sin a & AE/L\sin^2 a + 3EI/L^3\cos^2 a & 0 \\ 0 & 0 & 0 \end{bmatrix}$$

Note: $A = A_{ij}$

$$J_{jj} = \begin{bmatrix} AE/L\cos^2 a + 3EI/L^3\sin^2 a & (AE/L - 3EI/L^3)\cos a \sin a & 3EI/L^2\sin a \\ (AE/L - 3EI/L^3)\cos a \sin a & AE/L\sin^2 a + 3EI/L^3\cos^2 a & -3EI/L^2\cos a \\ 3EI/L^2\sin a & -3EI/L^2\cos a & 3EI/L \end{bmatrix}$$

THE FIXED-ROLLER ELEMENT:

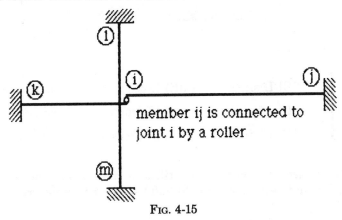

member ij is connected to
joint i by a roller

FIG. 4-15

The k matrices are the same as those for the fixed – pinned member
except that all the AE/L terms are set to zero.

	X_i	Y_i	θ_i	X_j	Y_j	θ_j
P_i	0	0	0	0	0	0
V_i	0	$3EI/L^3$	0	0	$3EI/L^3$	$3EI/L^2$
M_i	0	0	0	0	0	0
P_j	0	0	0	0	0	0
V_j	0	$3EI/L^3$	0	0	$3EI/L^3$	$3EI/L^2$
M_j	0	$3EI/L^2$	0	0	$3EI/L^2$	$3EI/L$

$$S_{ij} = \begin{bmatrix} -3EI/L^3\sin^2 a & 3EI/L^3\cos a \sin a & -3EI/L^2\sin a \\ 3EI/L^3\cos a \sin a & -3EI/L^3\cos^2 a & 3EI/L^2\cos a \\ 0 & 0 & 0 \end{bmatrix}$$

$$S_{ji} = \begin{bmatrix} -3EI/L^3\sin^2a & +3EI/L^3\cos a \sin a & 0 \\ 3EI/L^3\cos a \sin a & -3EI/L^3\cos^2a & 0 \\ -3EI/L^2\sin a & 3EI/L^2\cos a & 0 \end{bmatrix}$$

$$J_{ii} = \begin{bmatrix} 3EI/L^3\sin^2a & -3EI/L^3\cos a \sin a & 0 \\ -3EI/L^3\cos a \sin a & 3EI/L^3\cos^2a & 0 \\ 0 & 0 & 0 \end{bmatrix}$$

$$J_{jj} = \begin{bmatrix} 3EI/L^3\sin^2a & -3EI/L^3\cos a \sin a & 3EI/L^2\sin a \\ -3EI/L^3\cos a \sin a & 3EI/L^3\cos^2a & -3EI/L^2\cos a \\ 3EI/L^2\sin a & -3EI/L^2\cos a & 3EI/L \end{bmatrix}$$

Example Problem 4-6: Form the stiffness matrix for the structure shown. Evaluate all terms. $E = 30,000$ ksi (see Example Problem 4-3)

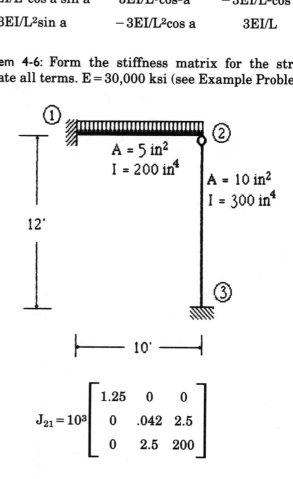

$$J_{21} = 10^3 \begin{bmatrix} 1.25 & 0 & 0 \\ 0 & .042 & 2.5 \\ 0 & 2.5 & 200 \end{bmatrix}$$

$$J_{23} = 10^3 \begin{bmatrix} .009 & 0 & 0 \\ 0 & 2.083 & 0 \\ 0 & 0 & 0 \end{bmatrix}$$

$$J_2 = 10^3 \begin{bmatrix} 1.26 & 0 & 0 \\ 0 & 2.13 & 2.5 \\ 0 & 2.5 & 200 \end{bmatrix}$$

Note the lack of symmetry created by the internal hinge.

Problems for Solution: For the structures shown in problems 4-1 thru 4-5, form the structural stiffness matrix and apply the boundary conditions.

4-1

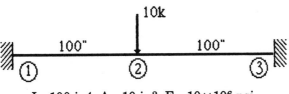

$$I = 100 \text{ in}^4, A = 10 \text{ in}^2, E = 10 \times 10^6 \text{ psi}$$

4-2

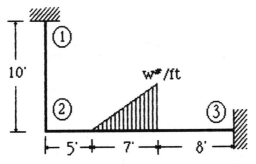

$A_{12} = 2.5 \text{ in}^2, I_{12} = 220 \text{ in}^4, E_{12} = 30000 \text{ ksi},$
$A_{23} = 5.0 \text{ in}^2, I_{23} = 560 \text{ in}^4, E_{23} = 22000 \text{ ksi}.$

4-3

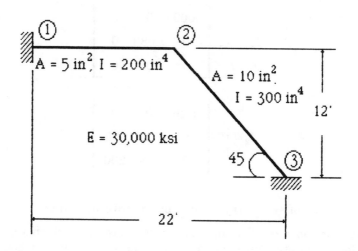

A = 5 in², I = 200 in⁴

A = 10 in²

I = 300 in⁴

12'

E = 30,000 ksi

45

22'

4-4

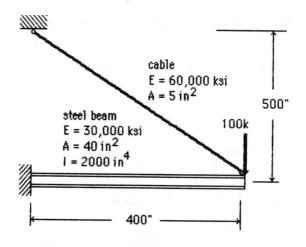

cable
E = 60,000 ksi
A = 5 in²

500"

steel beam
E = 30,000 ksi
A = 40 in²
I = 2000 in⁴

100k

400"

4-5

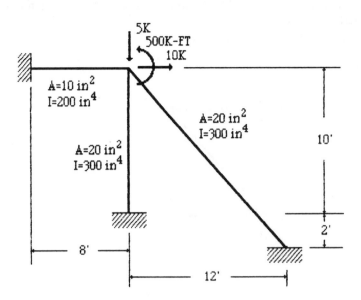

4-6. The shear wall shown is subjected to a wind loading. Discuss the effect on the math model of whether or not the beam element available includes the shear stiffness as well as the flexural stiffness. The roof acts as a diaphram delivering the wind loads on the end wall to the side walls as shear loads.

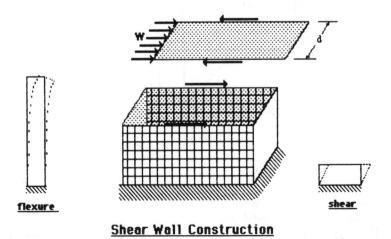

Shear Wall Construction

4-7: Form the stiffness matrix for the truss shown and apply the B.C.

All members have a cross-sectional area $A = 2.5$ in^2 and a modulus of elasticity $E = 30000$ ksi.

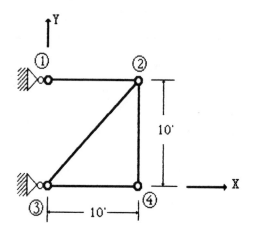

4-8: Form the stiffness matrix for the truss shown and apply the B.C.

All members have a cross-sectional area $A = 2.5$ in^2 and a modulus of elasticity $E = 30000$ ksi.

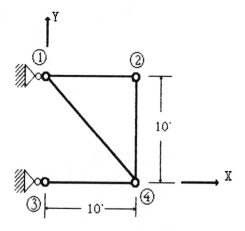

4-9: For the structure shown, form the structural stiffness matrix and apply the boundary conditions.

All members have a cross-sectional area $A = 2.5$ in^2 and a modulus of elasticity $E = 30000$ ksi.

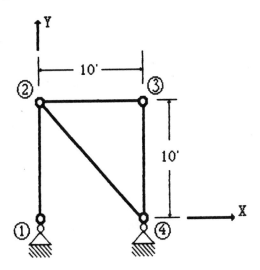

4-10: Form the stiffness matrix for the truss shown and apply the boundary conditions.

All members have a cross-sectional area $A = 2.5$ in^2 and a modulus of elasticity $E = 30000$ ksi.

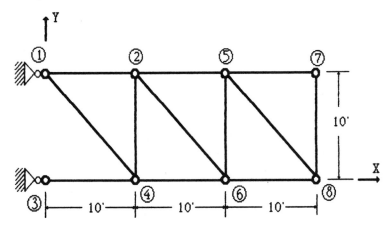

4-11: Elastic supports may be modeled by choosing the appropriate element to represent an elastic foundation. For instance, a translational spring support may be replaced by a pin-ended member. Develop the stiffness matrix for a rotational spring support.

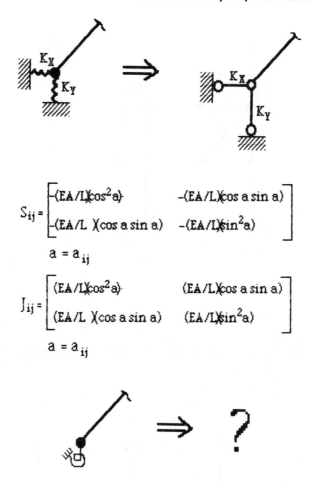

$$S_{ij} = \begin{bmatrix} -(EA/L)\cos^2 a) & -(EA/L)\cos a \sin a) \\ -(EA/L)(\cos a \sin a) & -(EA/L)\sin^2 a) \end{bmatrix}$$

$$a = a_{ij}$$

$$J_{ij} = \begin{bmatrix} (EA/L)\cos^2 a) & (EA/L)\cos a \sin a) \\ (EA/L)(\cos a \sin a) & (EA/L)\sin^2 a) \end{bmatrix}$$

$$a = a_{ij}$$

4-12. Set up the stiffness matrix for the structure shown and apply the boundary conditions. $E = 30,000$ ski

member	A (in²)	I (in⁴)
12	10	100
56	10	100
23	15	200
35	15	200
24	12	
45	12	
34	12	

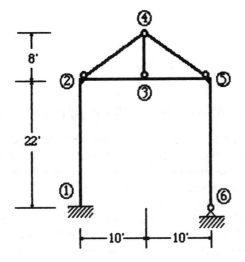

CHAPTER 5

THE LOAD AND DISPLACEMENT VECTORS

In the equation $X = K^{-1} P$, P is the vector of all the external forces acting on the joints of the structure, *including the unknown reactions*. If the structure is to be subjected to more than one loading condition†, then there will be more than one solution vector, ie., the order of the solution vector is the same as the order of the load vector.. Remembering that in matrix algebra an n x n times a n x m equals an n x m, or the number of rows of the solution vector is equal to the number of rows of the stiffness matrix and the number of columns of the solution vector is equal to the number of columns of the load vector, or , there is a column of displacements in the solution vector corresponding to each column of loads in the load vector. In general, a structure may be subjected to several loading conditions.

If a structure with n joints is subjected to three separate loading conditions. The equation $P = KX$ is:

$$
\begin{Bmatrix}
\overset{n \times 3}{} \\
\begin{array}{ccc}
P & P & P \\
V & V & V \\
M & M & M \\
\hline
P & P & P \\
V & V & V \\
M & M & M \\
\vdots & \vdots & \vdots \\
\hline
P & P & P \\
V & V & V \\
M & M & M
\end{array}
\end{Bmatrix}
\begin{matrix} 1 \\ \\ \\ 2 \\ \\ \\ \\ \\ n \end{matrix}
=
\begin{bmatrix} \overset{n \times n}{\mathbf{K}} \end{bmatrix}
\begin{Bmatrix}
\overset{n \times 3}{} \\
\begin{array}{ccc}
X & X & X \\
Y & Y & Y \\
\theta & \theta & \theta \\
\hline
X & X & X \\
Y & Y & Y \\
\theta & \theta & \theta \\
\vdots & \vdots & \vdots \\
\hline
X & X & X \\
Y & Y & Y \\
\theta & \theta & \theta
\end{array}
\end{Bmatrix}
\begin{matrix} 1 \\ \\ \\ 2 \\ \\ \\ \\ \\ n \end{matrix}
$$

†A loading condition is all the loads applied to a structure at a given time, under a given set of circumstances.

Note that there is a column vector for each loading condition and a solution vector corresponding to each loading condition.

Types of loads: Loads are broadly categorized as dead loads, live loads, environmental loads, fabrication loads, or operating loads. It is a matter of judgement into which category a particular load is placed. Some loads are static and some loads are dynamic.

Dead loads include the self weight of the structure as well as the weight of all items permanently attached to the structure.

Live loads include those loads which are not permanently acting on the structure, but vary with time. Examples of live loads are people, movable partitions, vehicles, overhead cranes, equipment, machinery, etc.

Environmental loads include wind, snow and ice loads, rainwater ponding, earthquake loads, temperature loads, blast loads, wave and current loads, hydrostatic pressure, earth pressures, support settlement, etc.

Fabrication loads take into account installation forces, residual forces, handling forces, pre-stressing and post-tensioning, etc.

Operating loads include vehicle loads, impact, overhead cranes, machinery and equipment loads, etc.

An example of a typical vehicle load is a twenty ton truck (Fig. 5-1).

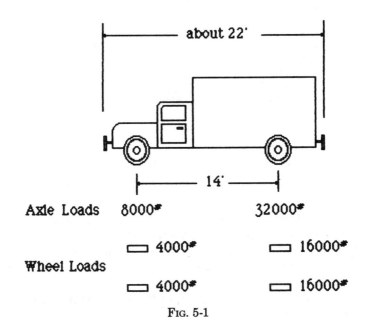

FIG. 5-1

Earthquake loads and wind loads act laterally on a structure. Earthquake loads are inertial loads resulting from motion of the base of the structure. Wind loads, on the other hand, are applied to the projected area of the structure. However, it is common practice to distribute both of these loads to the joints of the structure according to methods dictated by the applicable specification.

Wind loads are usually specified in terms of static pressure as a function of the height above grade. The pressure loading is then converted into equivalent joint loads by simple proportioning and applied at the roof and at each floor level. Wind loads may also be applied to the roof, normal to the surface.

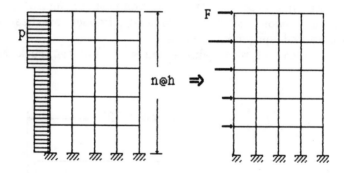

FIG. 5-2. *Wind Loads*

Earthquake loads are generaly considered to be the inertial forces resulting from ground motions. The inertial forces may be determined by a dynamic analisis, lumping the mass of the structure at selected nodes. A more common method is to calculate, from a code formula, a value for the total shear, V, at the base of the structure and then to distribute lateral forces equivalent to the base shear to the joints at the roof and at each floor level.

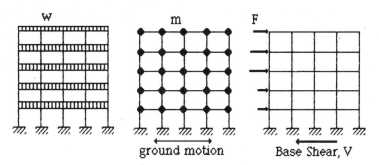

FIG. 5-3. *Earthquake Loads*

It has been shown by dynamic analysis that the maximum acceleration of a simple mass-spring system is proportional to the natural frequency of the system and to the maximum base velocity. This in turn is converted to show that the maximum response acceleration is proportional to the acceleration due to gravity, or $a_{max} = N\,g$, where N is defined as a shock factor and g is the gravitational acceleration. The equivalent static force is then calculated as $V = m\;a_{max} = W/g(N\;g) = NW$. W is the static weight of the structure.

Factors affecting the determination of the "equivalent" static load are:

> Nature of occupancy and use
> Overall height
> Height to width ratio
> Location and topography of the site
> Nature of the soil
> Dead loads
> Structural stiffness and type of framing system
> Distribution of the structural mass and stiffness
> The shape of the structural form

LOAD COMBINATIONS:

In the design codes and building codes, various loadings and combinations of loadings are specified. The intent is to find the most severe loading condition for each and every member in the structure. It may be, for instance, that the worst loading condition for the roof members is combined dead loads and wind loads, while for some other region of the structure such as the floor joists, the combination of dead loads and live loads may be the most severe case. The distribution of loads must also be considered. (See Example 5-3)

A specification might state that the following loadings are to be considered.

> D : dead loads
> L : live loads
> L_r : roof live load
> W : wind load
> S : snow load
> E : earthquake
>   load
> R : rain or ice loads

The loads, multiplied by their appropriate load factors, are to be applied in the following combinations, as separate loading conditions.

> 1.4 D
> 1.2 D + 1.6 L + 0.5(L_r or S or R)
> 1.2 D + 1.6 (L_r or S or R) + (0.5 L or 0.8 W)
> 1.2 D + 1.3W + 0.5 L + 0.5((L_r or S or R)
> 1.2 D + 1.5 E + (0.5 L or 0.2 S)
> 0.9 D − (1.3W or 1.5 E)

The load factors are statistically arrived at by considering the degree of certainty of each type of load and the probability of these loads occuring in combination.

If only the six loading conditions listed above are investigated, the load vector will have six columns and, therefore, the solution vector of joint displacements will have six columns. Each column of the load vector represents a separate loading condition and the corresponding column of the solution vector represents the joint displacements caused by that loading condition. However, within each of the specified loading conditions, the distribution of loads causing the worst case for each part of the structure must be investigated. Consequently, there may be a dozen or more significant loading conditions.

Example 5.1 The building structure shown has uniform dead load distribution.

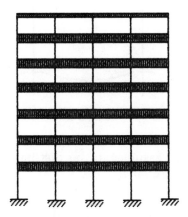

FIG. 5-4. *Dead Load Distribution*

If the live load is assumed to be uniformly distributed as shown in Fig.(5-5),

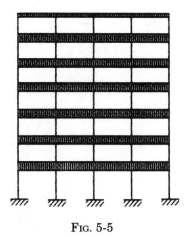

FIG. 5-5

the displaced shape might look as shown in Fig. 5-6.

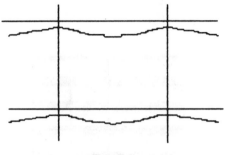

FIG. 5-6

If the live load is assumed to be distibuted in a checkerboard manner, as shown in Fig. 5-7,

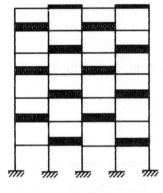

FIG. 5-7

the displaced shape might look as shown in Fig. 5-8.

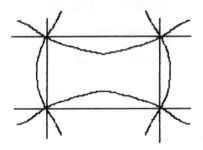

FIG. 5-8

In the uniformly distributed live load case, the columns and beams are not rotated at the ends and therefore the columns essentially see axial loads only (except for the outside columns). For the checkerboard distribution, the columns and beams rotate at the ends relieving the negative end moments and increasing the positive moments in the center of the beams and increasing the effective length of the columns. The effects of the assumed distribution of the live load are a function of the relative value of the dead load to the live load. If the dead load is large compared to the live load then the distribution of the live load is less significant.

Sign Convention for External Forces: The externally applied loads, reactions, boundary conditions and resulting displacements are in the the global system.

For all the problems in this book, the global system is:

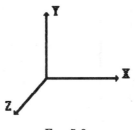

Fig. 5-9

Forces or translations are positive when directed in the positive X, Y, or Z directions. For moments, the right-hand rule is used. Place your thumb in the direction of the positive X, Y, or Z axis and your fingers will curl in the direction of a positive moment or rotation.

EXAMPLE 5.2:

Set up the equation $P = KX$ for the structure shown in Fig. 5-10 for the loading combinations previously specified. The dead weight (ignoring the weight of the members) is D# and there is a wind load acting at joint 4 of W#. K is in #/ft and, therefore, the joint displacements will be in feet.

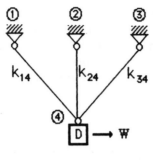

FIG. 5-10

Solution:

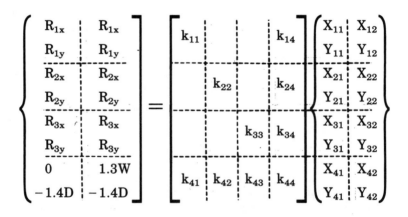

The subscript notation for the joint forces are first the joint number and second the direction in the Global System. Each column represents a unique loading condition.

The double subscript notation on the joint diplacements indicate first the joint number and second the loading condition number. The values in column one represent the joint displacements for loading condition one and the values in column two represent the displacements for loading condition two.

In linear elastic analyses, the displacements are a linear function of the loads. When there are multiple combinations of the various loads, it is sometimes more efficient to analyse the structure for 1.0 D, again for 1.0 L, again for 1.0 W, etc. and then to superpose the results, multiplying the individual results by the appropriate load factors.

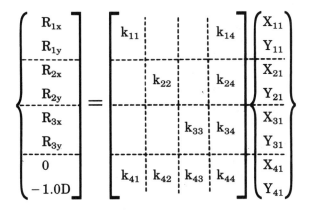

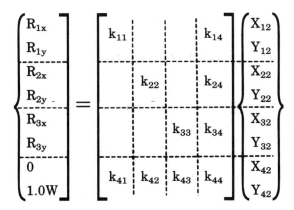

$$X^1{}_{ij} = 1.4X_{i1} , \quad X^2{}_{ij} = 1.2X_{i1} + 1.3 \, X_{i2}$$

$$Y^1{}_{ij} = 1.4Y_{i1} , \quad Y^2{}_{ij} = 1.2Y_{i1} + 1.3 \, Y_{i2}$$

Here the superscripts indicate the particular loading condition with appropriate load factors.

In this way the structure need be analysed only once for each type of loading and the results used as many times as necessary, multiplied by the applicable load factors.

Should a given load type change value, it is not necessary to rerun the computer analysis, but simply to ratio the results for that particular load type and to redo the superposition of effects.

In many structural computer programs, a dead weight analysis can be performed automatically. The input data for joint coordinates, member connectivity, and cross-sectional area previously input are used to calculate the volume of the members. The material property of weight densities is all the additional data the program requires to calculate the dead weight.

MEMBER FORCES:

In the problems in previous chapters, the structures were loaded at the joints only. In reality, loads are most often applied to the members. The commonly accepted exception to this, are truss loads. By definition, trusses are loaded only at the joints as it is presumed that the members carry only axial loads. However, while the self weight of the truss members and the wind forces on them may be small and of secondary importance, the dynamic inertial forces may not be and must be taken into consideration.

Many computer programs allow member loads to be input directly. In this book, the loads must be applied directly to the joints, the member forces must be replaced by statically equivalent joint forces which result in the same joint displacements as those caused by the real element loads (See Ex. 5-3). Subsequent to the solution using joint loads, to determine the inernal forces in an element, the individual element is removed as a free body with all the external forces acting on it. These forces are the member forces due to the joint loads plus the original member loads. The final results are obtained by use of the equilibrium equations.

Example 5-3: In the structure shown in Fig. 5-11, there are member forces on members 12 and 23. Determine the equivalent joint forces. (Refer to App. D)

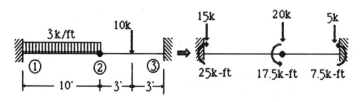

FIG. 5-11

Problem 5.1: Set up the load vector for the structure shown. The specification calls for the following loading conditions: (1.4D) and (1.2D + 1.6L). The live load is shown as 200#/ft². The concrete deck weighs 150#/ft³ and is 8 inches thick. The beams and trusses are steel with a weight density of 284#/in³. The beams are W10x54. The truss members have a cross-sectional area of 2.5 in².

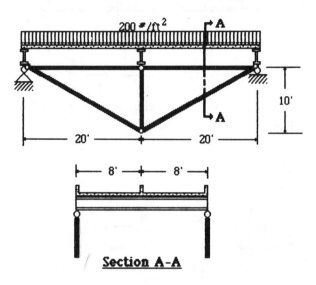

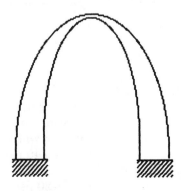

Section A-A

Problem 5.2: Discuss the various loads that may be applied to the St. Louis Arch and discuss how you would set up the load vector. There is an elevator inside the arch which takes people to the top.

5.3: Evaluate the joint loads which are equivalent to the loads shown.

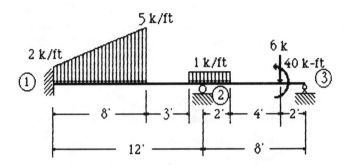

5-4: Set up the load vector using the load factors and combinations referred to previously in this chapter.

Lateral Loads:

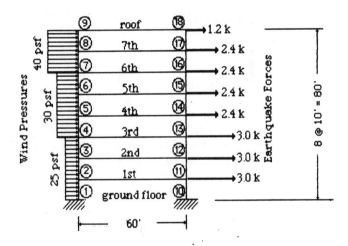

Vertical Loads:

Dead Loads:
Roof—12 psf
Exterior walls—20 psf
Interior Walls—10 psf
Floors—40 psf

Live Loads:
Roof—20 psf
Floors—50 psf

5-5: Determine the joint loads equivalent to a line of twenty ton trucks bumper-to-bumper on the bridge. (see Fig. 5-1) The Piers are spaced at 150'.

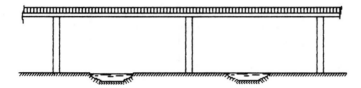

5-6: If there is a single twenty ton truck on the bridge, determine the location of the truck to cause the worst case loading for a) shear and b) moment in the bridge deck.

5-7: Set up the load vector for an analysis of the bridge deck for the loading conditions of problems 5-5 and 5-6 in addition to dead load. The weight of the reinforced concrete may be taken as 150#/ft^3.

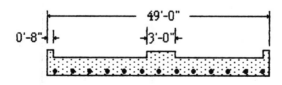

Bridge Deck Cross-Section

5-8: For the building frame shown here, discuss the worst loadings condition for each member of the structure.

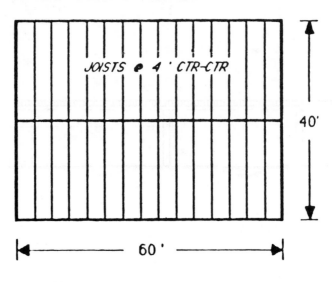

PLAN VIEW

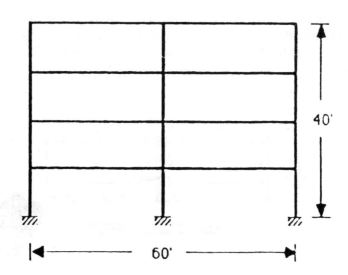

N/S ELEVATION
(E/W ELEVATION SIMILAR)

STRUCTURAL IDEALIZATION

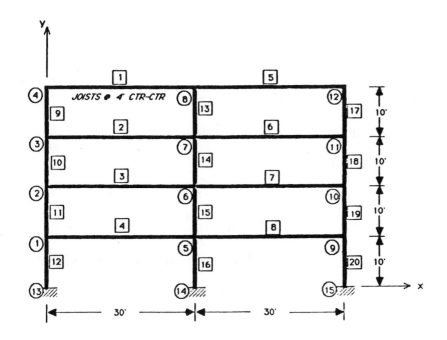

N/S ELEVATION

DEFINITION OF LOADS

	ROOF	FLOORS	WALLS
DEAD LOADS (PSF)	10	20	15
LIVE LOADS (PSF)	20	40	
WIND LOADS (PSF)	−24		30 (0–30′ ELEV)
			36 (30–40′ ELEV)

LOADING CONDITIONS

1) $1.4D + 1.7L$

2) $0.75\,(1.4D + 0.7L + 1.7W)$

MEMBER PROPERTIES

MEMBER	SECTION	A, in^2	I, in^4
1, 5	W12×26	7.65	204
2, 3, 4, 6, 7, 8	W14×34	10.0	340
9, 17, & JOISTS	W6×15	4.56	9.67
10, 13, 18	W6×25	7.34	17.1
11, 14, 19	W8×35	10.3	42.6
12, 15, 20	W10×45	13.3	53.4
16	W10×60	17.7	116

E = 30,000 ksi FOR ALL MEMBERS

CHAPTER 6

ANALYSING PLANE FRAMES AND TRUSSES

PROCEDURE FOR FORMING THE STRUCTURAL STIFFNESS MATRIX (FROM CHAPTER 4)

1. Draw the idealized structure, numbering all joints and members.

2. Establish the Global Axes.

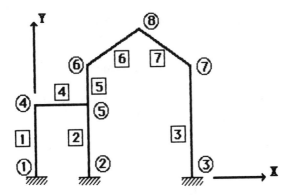

3. Set up the pattern of the stiffness matrix.

$$
K = \begin{array}{|c|c|c|c|c|c|c|c|}
\hline
J_1 & & & S_{14} & & & & \\
\hline
& J_2 & & & S_{25} & & & \\
\hline
& & J_3 & & & & S_{37} & \\
\hline
S_{41} & & & J_4 & S_{45} & & & \\
\hline
& S_{52} & & S_{54} & J_5 & S_{56} & & \\
\hline
\multicolumn{3}{|c|}{\text{Symmetric}} & & S_{65} & J_6 & & S_{68} \\
\hline
& & S_{73} & & & & J_7 & S_{78} \\
\hline
& & & & & S_{86} & S_{87} & J_8 \\
\hline
\end{array}
$$

4. Calculate each member stiffness, S_{ij}, for i<j, and locate the result in its proper place in the stiffness matrix.

$$S_{ij} = a^t_{ij} \, k_{ij} \, a_{ji}$$

5. Determine each S_{ij}, for i<j, by taking the corresponding transpose and locate the result in its proper place in the stiffness matrix below the diagonal.

$$S_{ji} = S^t_{ij}, \; i<j$$

6. Sum all the S_{ij} sub-matrices in each row and place the sum on the main diagonal, modifying the sum by:
   (A) Multply by (-1) the first and second *columns*.
   (B) Multiply by (2) the last element in the third row.
   The result is a joint stiffness matrix.
   Note that the joint stiffness matrix J_i equals the sum of the modified member stiffness matrices for members attached to that joint, ie, sum across the rows and modify the result.

$$J_i = \Sigma J_{in} = [\Sigma a^t_{ij} \, k_{ii} \, a_{ij} = \Sigma a^t_{ij} k_{ij} \, a_{ji}] \text{ (modified)} = \Sigma S_{ij} \text{ (modified)}$$

Now that the stiffness matrix has been formed, the solution vector may be obtained by proceding as follows:

7. Apply the boundary conditions.

$$K =$$

J_4	S_{45}			
S_{54}	J_5	S_{56}		
	S_{65}	J_6		S_{68}
	Symm.		J_7	S_{78}
		S_{86}	S_{87}	J_8

8. Form the load vector, P (Externally applied joint forces in the global system):
   (A) Calculate the fixed end forces for each member and transfer them to the global coordinate system.

(B) Add all the fixed end forces at each joint.
(C) Add the sum to the loads applied directly to the joints.
(D) Arrange the forces in the same order as the terms in the stiffness matrix.
(E) Repeat for each loading condition.

9. Set up the final equation to be solved: $P = KD$

$$
\begin{pmatrix}
\dfrac{\begin{matrix} F_X \\ F_Y \\ M \end{matrix}}{\begin{matrix} F_X \\ F_Y \\ M \end{matrix}}_{\!\!1} \\[2pt]
\dfrac{\begin{matrix} F_X \\ F_Y \\ M \end{matrix}}{\begin{matrix} F_X \\ F_Y \\ M \end{matrix}}_{\!\!j} \\[2pt]
\begin{matrix} F_X \\ F_Y \\ M \end{matrix}_{\!\!k}
\end{pmatrix}
=
\begin{bmatrix}
[J]_i & [S]_{ij} & [S]_{ik} \\
[S]_{ji} & [J]_j & [S]_{jk} \\
[S]_{ki} & [S]_{kj} & [J]_k
\end{bmatrix}
\begin{pmatrix}
\begin{matrix} X \\ Y \\ \theta \end{matrix}_{\!\!i} \\
\begin{matrix} X \\ Y \\ \theta \end{matrix}_{\!\!j} \\
\begin{matrix} X \\ Y \\ \theta \end{matrix}_{\!\!k}
\end{pmatrix}
$$

10. Solve for the joint displacements for each loading condition.

$$D = K^{-1}P$$

11. Calculate the member end forces in the local coordinate systems (as developed in Chapter 4).

$$f_{ij} = k_{ii} \, a_{ij} \, D_i + k_{ij} \, a_{ji} \, D_j$$
$$f_{ji} = k_{ji} \, a_{ij} \, D_i + k_{jj} \, a_{ji} \, D_j$$

12. For each member, add algebraically the fixed end forces (as calculated in step 8A) to the forces calculated in step 11. (See Ex. 6-1.)

13. Draw free body diagrams and check equilibrium.

14. Plot the displaced shape.

Example Problem 6-1 For the structure shown, determine the forces in member 12 and plot the displaced shape. [See Example Problem (4-3)]
$E = 30,000$ ksi

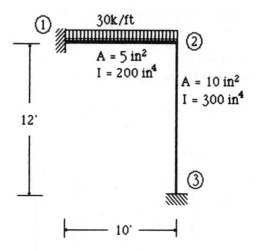

From Example Problem 4-3:

$$K = 10^3 \begin{bmatrix} 1.286 & 0 & -2.6 \\ 0 & 2.13 & 2.5 \\ -2.6 & 2.5 & 450.0 \end{bmatrix}$$

Equivalent joint loads:

wL/2 = 150 k   wL/2

w k/ft

L

$wL^2/12$ = 3000 k-in $wL^2/12$

Set up the load vector:

$$\begin{Bmatrix} 0 \\ -150 \\ -3000 \end{Bmatrix}$$

Write the equation, $P = KX$:

$$\begin{Bmatrix} 0 \\ -150 \\ -3000 \end{Bmatrix} = 10^3 \begin{bmatrix} 1.286 & 0 & -2.6 \\ 0 & 2.13 & 2.5 \\ -2.6 & 2.5 & 450.0 \end{bmatrix} \begin{Bmatrix} X \\ Y \\ Z \end{Bmatrix}_2$$

Solve for the displacement vector:

$$10^3 \begin{Bmatrix} X \\ Y \\ \theta \end{Bmatrix}_2 = \begin{Bmatrix} -1.0769 \\ -5.2434 \\ -0.53265 \end{Bmatrix}_2 , \quad \begin{matrix} X_2 = -0.013'' \\ Y_2 = -0.063'' \\ \theta_2 = -0.00639 \text{ rad.} \end{matrix}$$

Displaced Shape

Solve for member forces:

$$f_{21} = k_{22} \, a_{21} \, D_2 + k_{21} \, a_{12} \, D_1$$

$$f_{21} = 10^3 \begin{bmatrix} 1.25 & 0 & 0 \\ 0 & .042 & 2.5 \\ 0 & 2.5 & 200 \end{bmatrix} \begin{bmatrix} 1 & 0 & 0 \\ 0 & 1 & 0 \\ 0 & 0 & 1 \end{bmatrix} \begin{Bmatrix} -.013 \\ -.063 \\ -.00639 \end{Bmatrix}$$

$P_{21} = 10^3(1.25)(-.013) = -16.25 \text{ k}$

$V_{21} = 10^3[(.042)(-.063) + (2.5)(-.00639)] = -18.62 \text{ k}$

$M_{21} = 10^3[(2.5)(-.063) + (200)(-.00639)] = -1435.5 \text{ k-in} = -119.6 \text{ k-ft}$

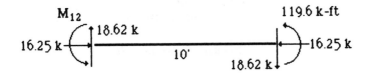

$$\Sigma M_1 = 0: \ -M_{12} + (18.62)(10) - 119.6 = 0, \ M_{12} = 66.6 \text{ k-ft}$$

Adding the member forces due to joint loads to the fixed-end forces due to member loads gives the member forces due to all the loads on the structure.

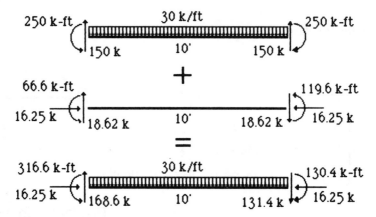

Example Problem 6-2: Determine the member forces for the structure shown. There are two loading conditions. (1) D.L. = 0.6 k/ft (2) L.L. as shown in sketch.

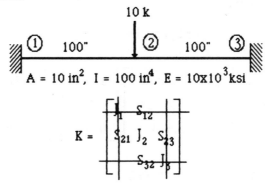

In this case, the S matrices are not required, only J_2 is required. Therefore J_{21} & J_{23} are evaluated and added to get J_2.

$$EA/L = 1 \times 10^3 \text{ k/in}$$
$$12EI/L^3 = 0.012 \times 10^3 \text{ k/in}$$
$$6EI/L^2 = 0.6 \times 10^3 \text{ k/in}$$
$$4EI/L = 40 \times 10^3 \text{ k/in}$$

$$\cos a_{21} = 1, \ \sin a_{21} = 0$$

$$J_{21} = 10^3 \begin{bmatrix} 1 & 0 & 0 \\ 0 & .012 & 0.6 \\ 0 & 0.6 & 40 \end{bmatrix}$$

$$\cos a_{23} = -1, \sin a_{23} = 0$$

$$J_{23} = 10^3 \begin{bmatrix} 1 & 0 & 0 \\ 0 & .012 & -0.6 \\ 0 & -0.6 & 40 \end{bmatrix}$$

$$J_2 = J_{21} + J_{23} = 10^3 \begin{bmatrix} 2 & 0 & 0 \\ 0 & 0.024 & 0 \\ 0 & 0 & 80 \end{bmatrix}$$

Loading Condition 1:

$$\begin{Bmatrix} 0 \\ -10 \\ 0 \end{Bmatrix} = 10^3 \begin{bmatrix} 2 & 0 & 0 \\ 0 & 0.024 & 0 \\ 0 & 0 & 80 \end{bmatrix} \begin{Bmatrix} X \\ Y \\ \theta \end{Bmatrix}$$

$$X = 0, Y = -0.417''. \ \theta = 0$$

$$f_{21} = k_{22} \, a_{21} \, D_2 + k_{21} \, a_{12} \, D_2, \ D_1 = 0$$

$$f_{21} = 10^3 \begin{bmatrix} 1 & 0 & 0 \\ 0 & .012 & 0.6 \\ 0 & 0.6 & 40 \end{bmatrix} \begin{bmatrix} 1 & 0 & 0 \\ 0 & 1 & 0 \\ 0 & 0 & 1 \end{bmatrix} \begin{Bmatrix} 0 \\ -0.417 \\ 0 \end{Bmatrix} = \begin{Bmatrix} 0 \\ -5 \\ -250 \end{Bmatrix}$$

By symmetyry:

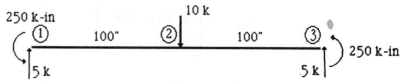

Loading Condition 2:

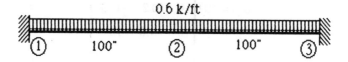

Equivalent Joint Loads:

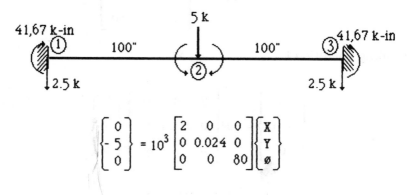

$$\left\{\begin{array}{c} 0 \\ -5 \\ 0 \end{array}\right\} = 10^3 \begin{bmatrix} 2 & 0 & 0 \\ 0 & 0.024 & 0 \\ 0 & 0 & 80 \end{bmatrix} \left\{\begin{array}{c} X \\ Y \\ \emptyset \end{array}\right\}$$

Since the second loading condition is half the first, the displacements and the member forces for the second loading condition will be half those of the first loading condition.

Member forces resulting from joint displacements caused by equivalent joint loads:

Fixed-end forces due to distributed loads:

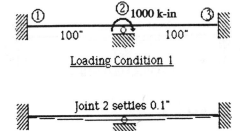

41.67 k-in   0.5 k/in   41.67 k-in   41.67 k-in 0.5 k/in   41.67 k-in

2.5 k   2.5 k   2.5 k   2.5 k

Member forces due to distributed loading:

166.7 k-in   0.5 k/in   83.3 k-in   83.3 k-in   0.5 k/in   166.7 k-in

5 k   5 k

Example 6-3: Determine the member forces for the two loading conditions shown.

①   ②1000 k-in   ③

100"   100"

Loading Condition 1

Joint 2 settles 0.1"

Loading Condition 2

The stiffness Matrix is:

$$J_2 = 10^3 \begin{bmatrix} 2 & 0 & 0 \\ 0 & .024 & 0 \\ 0 & 0 & 80 \end{bmatrix}$$

Loading Condition 1:

$$J_2 = 10^3 \begin{bmatrix} 2 & 0 & 0 \\ 0 & .024 & 0 \\ 0 & 0 & 80 \end{bmatrix}, \quad Y_2 = 0$$

$$\begin{Bmatrix} 0 \\ 1000 \end{Bmatrix} = 10^3 \begin{bmatrix} 2 & 0 \\ 0 & 80 \end{bmatrix} \begin{Bmatrix} X \\ \theta \end{Bmatrix}$$

$$X = 0, \ \theta = .0125 \text{ rad.}$$

$$M_{21} = 10^3(40)(.0125) = 500 \text{ k-in}$$

$$M_{23} = 10^3(40)(.0125) = 500 \text{ k-in}$$

Loading Condition 2:

$$Y_2 = -0.1''$$

$$P_2 = 0$$
$$V_2 = -2.4 \text{ k}$$
$$M_2 = 0$$

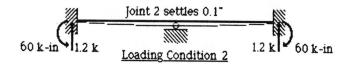

EXAMPLE PROBLEM 6-4:

Determine the displacements, reactions, and member forces for the structure shown.

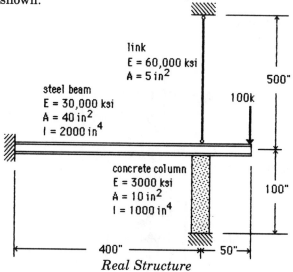

Real Structure

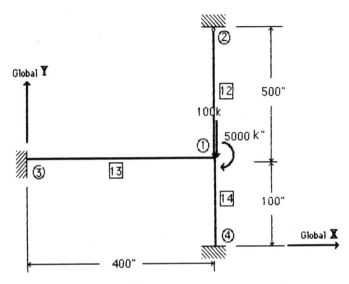

Idealized Structure

Establish the pattern of the stiffness matrix:

J_1	S_{12}	S_{13}	S_{14}
S_{21}	J_2		
S_{31}		J_3	
S_{41}			J_4

Apply the boundary conditions by crossing out the appropriate rows and corresponding columns. In this example, all the displacements at joints 2,3, and 4 are zero.

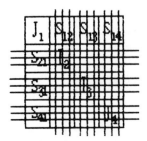

Calculate all S_{ij} sub-matrices above the main diagonal.

$$S_{ij} = \begin{bmatrix} -\left(\dfrac{EA}{L}\cos^2 a + \dfrac{12EI}{L^3}\sin^2 a\right) & -\left(\dfrac{EA}{L} - \dfrac{12EI}{L^3}\right)\cos a \sin a & -\dfrac{6EI}{L^2}\sin a \\[2ex] -\left(\dfrac{EA}{L} - \dfrac{12EI}{L^3}\right)\cos a \sin a & -\left(\dfrac{EA}{L}\sin^2 a + \dfrac{12EI}{L^3}\cos^2 a\right) & \dfrac{6EI}{L^2}\cos a \\[2ex] \dfrac{6EI}{L^2}\sin a & -\dfrac{6EI}{L^2}\cos a & \dfrac{2EI}{L} \end{bmatrix}$$

Member 1-2 is a link. A link cannot support shears or moments, therefore, all the shear and bending terms in the element stiffness matrix are set to zero.

$AE/L = 600 \text{ k/in}$

$$S_{12} = \begin{bmatrix} 0 & 0 & 0 \\ 0 & -600 & 0 \\ 0 & 0 & 0 \end{bmatrix}$$

② $\cos a = \dfrac{X_1 - X_2}{L_{12}} = 0$

$\sin a = \dfrac{Y_1 - Y_2}{L_{12}} = -1$

①   y →

↓ X

$AE/L = 3000 \text{ k/in}$
$12EI/L^3 = 11.25 \text{ k/in}$
$6EI/L^2 = 2250 \text{ k/in}$
$2EI/L = 3000{,}000 \text{ k/in}$

$$S_{13} = \begin{bmatrix} -3000 & 0 & 0 \\ 0 & -11.25 & 2250 \\ 0 & -2250 & 300{,}000 \end{bmatrix}, \quad$$

$\cos a = \dfrac{X_1 - X_3}{L_{13}} = 1$

$\sin a = \dfrac{Y_1 - Y_3}{L_{13}} = 0$

③   ① → x   (y ↑)

$$AE/L = 300 \text{ k/in}$$
$$12EI/L^3 = 36 \text{ k/in}$$
$$6EI/L^2 = 1800 \text{ k/in}$$
$$2EI/L = 60,000 \text{ k/in}$$

$$S_{14} = \begin{bmatrix} -36 & 0 & -1800 \\ 0 & -300 & 0 \\ 1800 & 0 & 60,000 \end{bmatrix}, \quad \begin{aligned} \cos a &= \frac{X_1 - X_4}{L_{14}} = 0 \\ \sin a &= \frac{Y_1 - Y_4}{L_{14}} = 1 \end{aligned}$$

Add all the sub-matrices S_{ij} in row one and locate the sum on the main diagonal. Modify the sum by multiplying columns 1 and 2 by (-1) and the last term by (2). Note, in this case only S_{12}, S_{13} and S_{14} need be calculated because the boundary conditions eliminate all the other S and J sub-matrices. (Note: In this case, it would be more efficient to calculate the J_{ii} matrices directly.)

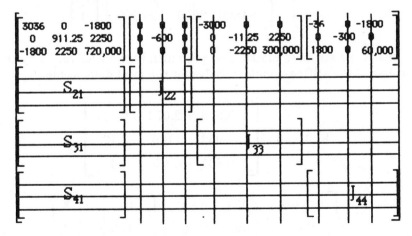

Calculate the joint loads and form the load vector.

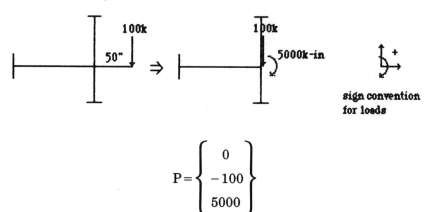

sign convention
for loads

$$P = \left\{ \begin{array}{c} 0 \\ -100 \\ 5000 \end{array} \right\}$$

The equation to be solved is:

$$\left\{ \begin{array}{c} 0 \\ -100 \\ 5000 \end{array} \right\} = \left[\begin{array}{ccc} 3036 & 0 & -1800 \\ 0 & 911.25 & 2250 \\ -1800 & 2250 & 720,000 \end{array} \right] \left\{ \begin{array}{c} X \\ Y \\ \theta \end{array} \right\}_1$$

Invert the stiffness matrix and premultiply the load vector by the inverse of the stiffness matrix. Solve for the joint displacements.

$$K^{-1} = 10^{-6} \left[\begin{array}{ccc} 329.8 & -2.052 & 0.831 \\ -2.052 & 1106 & -3.46 \\ 0.831 & -3.46 & 1.4 \end{array} \right]$$

$$10^{-6} \left[\begin{array}{ccc} 329.8 & -2.052 & 0.831 \\ -2.052 & 1106 & -3.46 \\ 0.831 & -3.46 & 1.4 \end{array} \right] \left\{ \begin{array}{c} 0 \\ -100 \\ 5000 \end{array} \right\} = \left\{ \begin{array}{c} X \\ Y \\ \theta \end{array} \right\}_1$$

$$\left\{ \begin{array}{c} X \\ Y \\ \theta \end{array} \right\}_1 = \left\{ \begin{array}{c} 0.00436 \\ -0.1279 \\ 0.00735 \end{array} \right\}$$

Calculate the member forces using;

$$f_{ij} = k_{ii} \, a_{ij} \, D_1, \, D_j = 0$$
$$f_{ji} = k_{ij} \, a_{ij} \, D_1, \, D_j = 0$$

or the forces at the far ends, f_{ji}, may be found by using the equilibrium equations for the element after determining the near end forces, f_{ij}.

$$f_{12} = \begin{bmatrix} 600 & 0 & 0 \\ 0 & 0 & 0 \\ 0 & 0 & 0 \end{bmatrix} \begin{bmatrix} 0 & -1 & 0 \\ 1 & 0 & 0 \\ 0 & 0 & 1 \end{bmatrix} \left\{ \begin{array}{c} 0.00436 \\ -0.1279 \\ 0.00735 \end{array} \right\} = \left\{ \begin{array}{c} 76.74 \\ 0 \\ 0 \end{array} \right\}$$

$$f_{13} = \begin{bmatrix} 3000 & 0 & 0 \\ 0 & 11.25 & 2250 \\ 0 & 2250 & 600,000 \end{bmatrix} \begin{bmatrix} 1 & 0 & 0 \\ 0 & 1 & 0 \\ 0 & 0 & 1 \end{bmatrix} \left\{ \begin{array}{c} 0.00436 \\ -0.1279 \\ 0.00735 \end{array} \right\} = \left\{ \begin{array}{c} 13.08 \\ 15.09 \\ 4120 \end{array} \right\}$$

$\Sigma F_x = 0$: $13.08 - P_3 = 0$, $P_3 = 13.08$ k
$\Sigma F_y = 0$: $15.09 - V_3 = 0$, $V_3 = 15.09$ k
$\Sigma M = 0$: $4120 - 15.09(400) + M_3 = 0$, $M_3 = 1916$ k-in

$$f_{14} = \begin{bmatrix} 300 & 0 & 0 \\ 0 & 36 & 1800 \\ 0 & 1800 & 120,000 \end{bmatrix} \begin{bmatrix} 0 & 1 & 0 \\ -1 & 0 & 0 \\ 0 & 0 & 1 \end{bmatrix} \left\{ \begin{array}{c} 0.00436 \\ -0.1279 \\ 0.00735 \end{array} \right\} = \left\{ \begin{array}{c} -38.37 \\ 13.07 \\ 873.7 \end{array} \right\}$$

$\Sigma M = 0$: $873.7 + M_4 - 13.07(100) = 0$: $M_4 = 433.3$ k-in

Verify equilibrium at joint (1):

$\Sigma Fx=0$: $13.07-13.08 = 0.01$, O.K.
$\Sigma FY=0$: $76.74-15.09+38.37-100 = 0.02$, O.K.
$\Sigma M =0$: $5000-4210-873.7 = 6.3$, O.K.

Verify equilibrium of entire structure:

$\Sigma F_x = 0$: $13.07 - 13.08 = 0$, o.k.
$\Sigma F_y = 0$: $-100 + 38.37 - 15.09 + 76.74 = 0$, o.k.
$\Sigma M = 0$: $5000 + 1916 + 433.3 - 13.07(100) - 15.09(400) = 0$: o.k.

In the Moment Distribution Method and the Slope Deflection Method, the link cannot be included in the analysis because the method considers only rotational stiffness and results in only the moments in the members and does not account for axial or shear forces:

If the moment distribution Method is used on this problem:
(D. F. = distribution factor = member stiffness divided by the joint stiffness, where joint stiffness = sum of the member stiffnesses at the joint.)

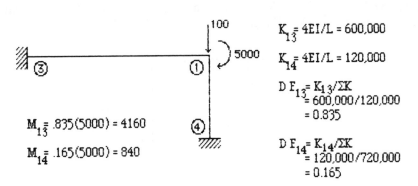

$$K_{13} = 4EI/L = 600,000$$

$$K_{14} = 4EI/L = 120,000$$

$$D F_{13} = K_{13}/\Sigma K \\ = 600,000/120,000 \\ = 0.835$$

$$M_{13} = .835(5000) = 4160$$

$$M_{14} = .165(5000) = 840$$

$$D F_{14} = K_{14}/\Sigma K \\ = 120,000/720,000 \\ = 0.165$$

To determine the axial and shear forces in the members, equilibrium equations must be used on the free body diagrams of the members.

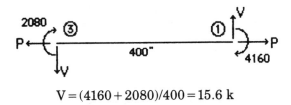

$$V = (4160 + 2080)/400 = 15.6 \text{ k}$$

The shear force in the beam increases to 15.6k which results in the axial force in the concrete column increasing from 38.37k to 84.4k while the axial force in the beam decreases. Note, also, the change in M4.

$$V = \frac{840 + 420}{100} = 12.6 \text{ k}$$

∴ The axial load in the beam, P = 12.6 k.

No method is exact, but some are better than others, depending on the circumstances. Matrix methods are generally best for complex structures.

Example Problem 6-5 Calculate the value of the member forces for the structure shown. (see Example Problem 4-6) E = 30,000 ksi

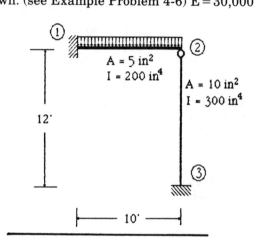

Fixed-End Forces:

$$
J_2 = 10^3 \begin{bmatrix} 1.26 & 0 & 0 \\ 0 & 2.13 & 2.5 \\ 0 & 2.5 & 200 \end{bmatrix}
$$

$$
\left\{ \begin{array}{c} 0 \\ -150 \\ -3000 \end{array} \right\} = 10^3 \begin{bmatrix} 1.26 & 0 & 0 \\ 0 & 2.13 & 2.5 \\ 0 & 2.5 & 200 \end{bmatrix} \left\{ \begin{array}{c} X \\ Y \\ \theta \end{array} \right\}_2
$$

$$X = -.000916''$$

$$Y = -.0535''$$

$$\theta = -.01432 \text{ rad.}$$

$$f_{21} = 10^3 \begin{bmatrix} 1.25 & 0 & 0 \\ 0 & .042 & 2.5 \\ 0 & 2.5 & 100 \end{bmatrix} \begin{Bmatrix} -.000916 \\ -.0535 \\ -.01432 \end{Bmatrix} = \begin{Bmatrix} -1.1 \text{ k} \\ -38 \text{ k} \\ -3000 \text{ k-in} \end{Bmatrix}$$

$$f_{23} = 10^3 \begin{bmatrix} 0 & 2.083 & 0 \\ -.009 & 0 & 0 \\ 0 & 0 & 0 \end{bmatrix} \begin{Bmatrix} -.000916 \\ -.0535 \\ -.01432 \end{Bmatrix} = \begin{Bmatrix} -111 \\ .008 \\ 0 \end{Bmatrix}$$

Check equilibrium of joint

$$\Sigma F_x = 0: 1.1 + .008 = 1.108 = 0$$

$$\Sigma F_y = 0: 38 + 111 - 150 = 0$$

$$\Sigma M = 0: 3000 - 3000 = 0$$

Example Problem 6-6 Determine the internal member forces and the external reactions for the structure shown. Use the same information given in Example Problem 4-4.

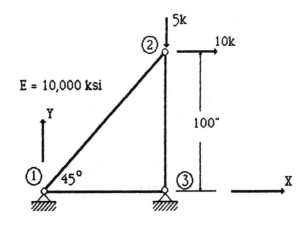

$$\begin{Bmatrix} 10 \\ -5 \end{Bmatrix} = 50 \begin{bmatrix} 1 & 1 \\ 1 & 5 \end{bmatrix} \begin{Bmatrix} X \\ Y \end{Bmatrix}_2$$

$$50 \begin{bmatrix} 1 & 1 \\ 1 & 5 \end{bmatrix}^{-1} \begin{Bmatrix} 10 \\ -5 \end{Bmatrix} = \begin{Bmatrix} X \\ Y \end{Bmatrix}_2$$

$$\begin{vmatrix} 50 & 50 \\ 50 & 250 \end{vmatrix} = 10,000$$

$$\frac{1}{10,000} \begin{vmatrix} 250 & -50 \\ -50 & 50 \end{vmatrix} \begin{Bmatrix} 10 \\ -5 \end{Bmatrix} = \begin{Bmatrix} X \\ Y \end{Bmatrix}_2$$

$$X = 0.275''$$

$$Y = -0.075''$$

Member Forces:

$$f_{ij} = k_{ii} \, a_{ij} \, D_i + k_{ij} \, a_{ji} \, D_j$$

(NOTE: $f_{13} = 0$ because $D_1 = D_3 = 0$.)

$$f_{12} = 0 + k_{12} \, a_{2i} \, D_2, \quad D_1 = 0$$

$\cos a_{21} = .707$, $\sin a_{21} = .707$

$$f_{12} = \begin{bmatrix} 100 & 0 \\ 0 & 0 \end{bmatrix} \begin{bmatrix} .707 & .707 \\ -.707 & .707 \end{bmatrix} \begin{Bmatrix} .275 \\ -.075 \end{Bmatrix} = \begin{Bmatrix} 14.14 \\ 0 \end{Bmatrix}$$

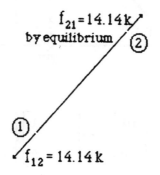

$$f_{23} = k_{22} \, a_{23} \, D_2 + k_{23} \, a_{32} \, D_3$$

$$f_{23} = k_{22} \, a_{23} \, D_2 + 0, \, D_3 = 0$$

$\cos a_{23} = 0, \, \sin a_{23} = 1$

$$f_{23} = \begin{bmatrix} -200 & 0 \\ 0 & 0 \end{bmatrix} \begin{bmatrix} 0 & 1 \\ -1 & 0 \end{bmatrix} \begin{Bmatrix} .275 \\ -.075 \end{Bmatrix} = \begin{Bmatrix} -15 \\ 0 \end{Bmatrix}$$

GLOBAL FORCES:

$$F_{12} = a_{12}^T \, f_{12} = \begin{bmatrix} -.707 & .707 \\ -.707 & -.707 \end{bmatrix} \begin{Bmatrix} 14.14 \\ 0 \end{Bmatrix} = \begin{Bmatrix} -10 \\ -10 \end{Bmatrix}$$

$$F_{23} = \begin{bmatrix} 0 & -1 \\ 1 & 0 \end{bmatrix} \begin{Bmatrix} -15 \\ 0 \end{Bmatrix} = \begin{Bmatrix} 0 \\ -15 \end{Bmatrix}$$

   The member forces are the forces acting on the members. The forces acting on a joint are equal and opposite to the algebraic sum of the member forces for the members attached to that joint.

Checking equilibrium at joint 2:

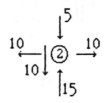

REACTIONS:

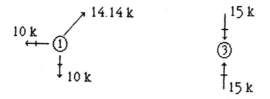

STRUCTURAL EQUILIBRIUM CHECK:

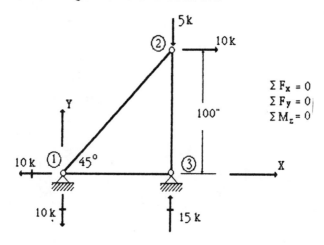

PROBLEMS FOR SOLUTION:

6-1 For the structure shown, determine the forces in the members caused by the live load shown and the dead load of the beam which weighs 100#/ft.

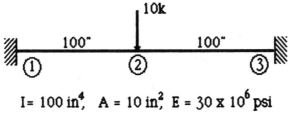

$I = 100$ in^4, $A = 10$ in^2, $E = 30$ x 10^6 psi

$I = 100$ in^4, A $= 10$ in^2, E $= 30 \times 10^6$ psi

6-2 Determine the member forces for the structure shown. $A_{12} = 2.5$ in^2, $I_{12} = 220$ in^4, $E_{12} = 30000$ ksi, $A_{23} = 5.0$ in^2, $I_{23} = 560$ in^4, $E_{23} = 22000$ ksi, $w = 1000 - $/ft.

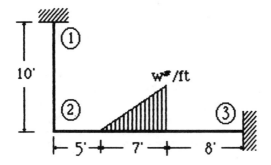

6-3 For the structure shown, determine the member forces caused by the dead weight of the structure and plot the displaced shape. The weight density of the material is 0.284#/in^3.

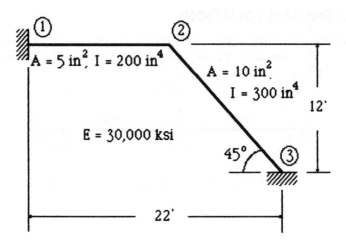

6-4 Determine the member forces for the structure shown.

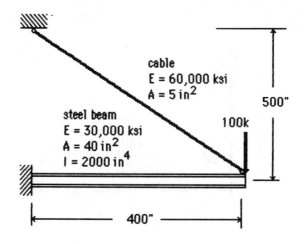

6-5 For the structure shown determine the forces in all the members for three loading conditions: 1 dead load only, 2 live load only, 3 combined dead and live loads. The forces shown are live loads. The structure is made of steel with a weight density of 0.284#/in³. E = 30,000 ksi

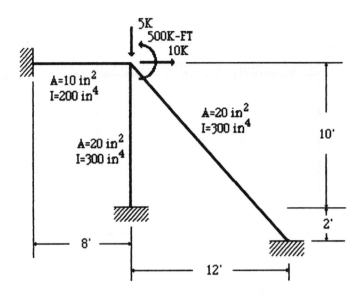

6-6 Determine the internal member forces and the external reactions for the structure shown. Compare the results with those in Chapter 3.

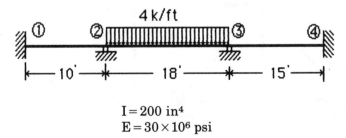

$$I = 200 \text{ in}^4$$
$$E = 30 \times 10^6 \text{ psi}$$

6-7 Determine the internal member forces, the external reactions and plot the displaced shape for the structure shown. Compare the results with those in Chapter 3. $EI = 30 \times 10^6$ psi, $I = 200$ in^4, $A = 10$ in^2

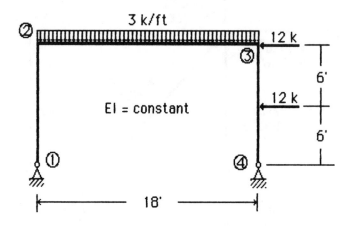

6-8 Determine the internal member forces, the external reactions and plot the displaced shape for the structure shown. Use the same information given in Example Problem 4-5.

Members 1–3, 3–4, & 2–3:
A = 10 in²
Members 1–2 & 2–4
A = 14.14 in²

L = 100″
E = 10 × 10³ ksi

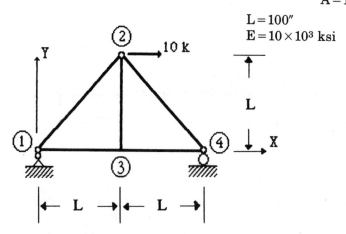

6-9 Determine the internal member forces for the structure shown. Refer to Example Problem 4-6. $E = 30 \times 10^3$ ksi

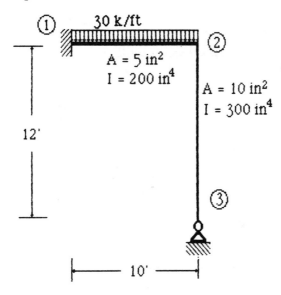

6-10 Compare the member forces of the structure of problem 6-9 with those of the structure shown.

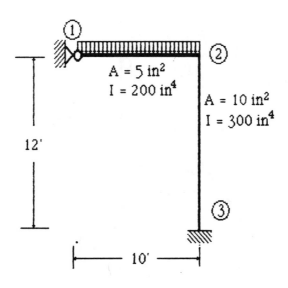

6-11 Find the value of the member forces for the structure shown.
E = 30,000 ksi

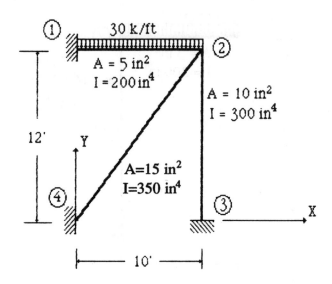

6-12 Find the value of the member forces for the structure shown.
E = 30,000 ksi

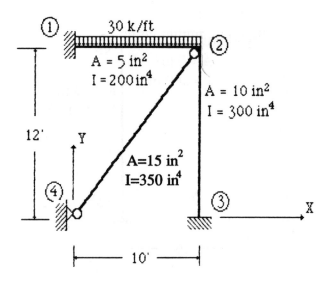

6-13 Find the value of the reactions on the beam which has the following properties. $E = 30,000$ ksi, $I = 200$ in⁴, $A = 10$ in².

Solve the following cases:

a) $K = 0$
b) $K = 0.5$ k/in
c) $K = 1.0$ k/in
d) $K = \infty$

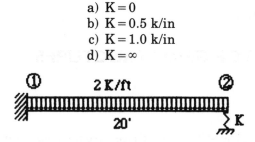

6-14 Using matrix methods, determine the value of the displacement of joint 1 and the value of the reactions for the structure shown.

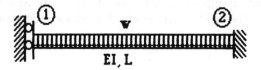

CHAPTER 7

ANALYSIS OF GRID STRUCTURES

In Chapter 4, the stiffness matrices are developed for planar structures which are loaded in the plane of the structure. There are, however, planar structures which are loaded perpendicular to the plane of the structure. These structural types are generally referred to as grids. Some examples of this type of structure are: roof and floor slabs, walls subjected to wind loads, & decks and gratings.

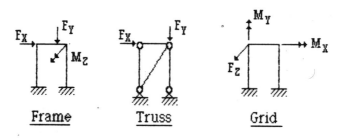

Fɪɢ. 7-1

For the plane frame and plane truss, the structure and the loads lie in the global X-Y plane. The degrees of freedom (joint displacements) are X, Y, and θ_Z. The member forces are longitudinal (axial) force in the local x direction, shear along the local z axis, and bending moment about the local z axis. For a plane grid the structure lies in the global X-Y plane, but the loads are out of plane, ie., in the global Z direction. The degrees of freedom are Z, θ_X, and θ_Y resulting in member forces of shear along the local z axis, bending moment about the local y axis, and torsional moment about the local x axis (Fig. 7-4).

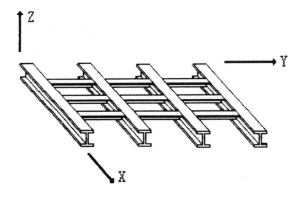

FIG. 7-2. *Grid Structure*

When the beams in one direction flex, the cross beams twist (Fig. 7-3).

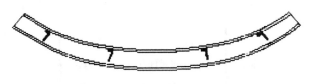

FIG. 7-3

Figure (7-4) shows the local and global coordinate systems for the grid element. The local y axis is referenced to the local x – global Y plane, except for vertical members, where it is referenced to the local x – global X plane.

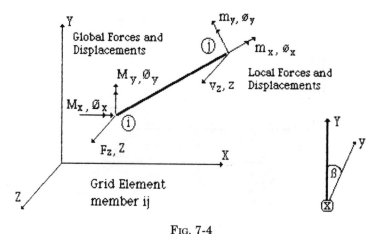

FIG. 7-4

The stiffness terms required for a grid are the shears and moments resulting from end displacements and the torque resulting from a torsional rotation.

$$M_x = (JG/L)\,(\theta x), \qquad m_y = (6EI/L^2)\,(z) + (4EI/L)\,(\theta y),$$
$$v_z = (12EI/L^3)\,(z) + (6EI/L^2)\,(\theta_y)$$

The procedure for forming the structural stiffness matrix for a grid structure is the same as that for a plane frame structure (see Ch. 4).

The k matrices are:

$$k_{ii} = \begin{bmatrix} JG/L & 0 & 0 \\ 0 & 4EIy/L & 6EIy/L^2 \\ 0 & 6EIy/L^2 & 12EIy/L^3 \end{bmatrix}$$

$$k_{ij} = \begin{bmatrix} JG/L & 0 & 0 \\ 0 & 2EIy/L & 6EIy/L^2 \\ 0 & 6EIy/L^2 & 12EIy/L^3 \end{bmatrix}$$

The formation of the S and J matrices follows the same procedure as for the plane frame element.

COORDINATE TRANSFORMATION:

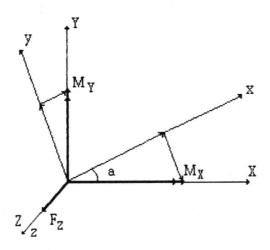

FIG. 7-5

$$m_x = M_x \cos a + M_Y \sin a$$
$$m_y = -M_x \sin a + M_Y \cos a$$
$$f_z = F_z$$

$$
\begin{Bmatrix} m_x \\ m_y \\ v_z \end{Bmatrix} =
\begin{bmatrix} \cos a & \sin a & 0 \\ -\sin a & \cos a & 0 \\ 0 & 0 & 1 \end{bmatrix}
\begin{Bmatrix} M_X \\ M_Y \\ F_Z \end{Bmatrix}
$$

or

$$
\begin{Bmatrix} \theta_x \\ \theta_y \\ z_z \end{Bmatrix} =
\begin{bmatrix} \cos a & \sin a & 0 \\ -\sin a & \cos a & 0 \\ 0 & 0 & 1 \end{bmatrix}
\begin{Bmatrix} \theta_X \\ \theta_Y \\ Z_Z \end{Bmatrix}
$$

Example Problem 7-1 For the grid structure shown, determine the member forces. $E = 30,000$ ksi

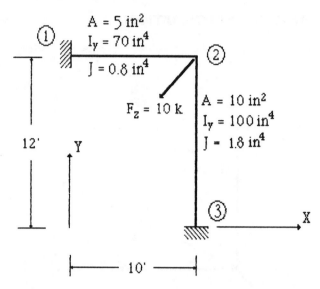

$G = E/[2(1+v)]$, v will be taken as 0.3 for all engineering materials in this text.

$$J_{ii} = \begin{bmatrix} \cos a_{ij} & \sin a_{ij} & 0 \\ \sin a_{ij} & \cos a_{ij} & 0 \\ 0 & 0 & 1 \end{bmatrix} E/L \begin{bmatrix} J/2.6 & 0 & 0 \\ 0 & 4I_y & 6I_y/L \\ 0 & 6I_y/L & 12I_y/L^2 \end{bmatrix} \begin{bmatrix} \cos a_{ij} & \sin a_{ij} & 0 \\ \sin a_{ij} & \cos a_{ij} & 0 \\ 0 & 0 & 1 \end{bmatrix}$$

$$J_{ii} = \frac{E}{L} \begin{bmatrix} J/2.6\cos a^2 + 4I_y \sin a^2 & (J/2.6 - 4I_y)\cos a_{ij} \sin a_{ij} & -6I_y/L \sin a_{ij} \\ (J/2.6 - 4I_y)\cos a_{ij} \sin a_{ij} & J/2.6 \sin a_{ij}^2 + 4I_y \cos a_{ij}^2 & 6I_y/L \cos a_{ij} \\ -6I_y/L \sin a_{ij} & 6I_y/L \cos a_{ij} & 12I_y/L^2 \end{bmatrix}$$

member 2-1: $E/L = 30000/120 = 250$, $J/2.6 = 0.8/2.6 = 0.308$,
$4I_y = 4(70) = 280$, $6I_y/L = 6(70)/120 = 3.5$,
$12I_y/L^2 = 12(70)1(120)^2 = 0.0583$
$\cos a_{21} = 1$· $\sin a_{21} = 0$

$$J_{22} = 250 \begin{bmatrix} 308 & 0 & 0 \\ 0 & 280 & 3.5 \\ 0 & 3.5 & .0583 \end{bmatrix}$$

member 2-3: $E/L = 30000/144 = 208$ $J/2.6 = 1.8/2.6 = 0.692,$
$4I_y = 4(100) = 400,$ $6I_{y/L} = 6(100)/144 = 4.17$
$12I_y/L^2 = 12(100)1(144)^2 = 0.0579$
$\cos a_{23} = 0, \sin a_{23} = 1$

$$J_{22} = 208 \begin{bmatrix} 400 & 0 & -4.17 \\ 0 & .692 & 0 \\ -4.17 & 0 & .0579 \end{bmatrix}$$

$$J_2 = 250 \begin{bmatrix} .308 & 0 & 0 \\ 0 & 280 & 3.5 \\ 0 & 3.5 & .0583 \end{bmatrix} + 208 \begin{bmatrix} 400 & 0 & -4.17 \\ 0 & .692 & 0 \\ -4.17 & 0 & .0579 \end{bmatrix}$$

$$J_2 = \begin{bmatrix} 83,277 & 0 & -867 \\ 0 & 70,144 & 875 \\ -867 & 875 & 26.6 \end{bmatrix}$$

$$\begin{Bmatrix} 0 \\ 0 \\ 10 \end{Bmatrix} = \begin{bmatrix} 83,277 & 0 & -867 \\ 0 & 70,144 & 875 \\ -867 & 875 & 26.6 \end{bmatrix} \begin{Bmatrix} \theta_x \\ \theta_y \\ Z \end{Bmatrix}$$

$$\begin{Bmatrix} \phi_x \\ \phi_y \\ Z \end{Bmatrix} = \begin{Bmatrix} .015636 \\ -.018734 \\ 1.5018 \end{Bmatrix}$$

$$f_{ij} = k_{ii} \, a_{ij} \, D_i$$

$$f_{21} = k_{22} \, a_{21} \, D_2$$

$$f_{21} = 250 \begin{bmatrix} .308 & 0 & 0 \\ 0 & 280 & 3.5 \\ 0 & 3.5 & .0583 \end{bmatrix} \begin{bmatrix} 1 & 0 & 0 \\ 0 & 1 & 0 \\ 0 & 0 & 1 \end{bmatrix} \begin{Bmatrix} .015636 \\ -.018734 \\ 1.5018 \end{Bmatrix}$$

$$f_{21} = \begin{Bmatrix} 1.2 \\ 2.7 \\ 5.5 \end{Bmatrix}$$

$$f_{23} = k_{22}\, a_{23}\, D_2$$

$$f_2 = 208 \begin{bmatrix} .692 & 0 & 0 \\ 0 & 400 & 4.17 \\ 0 & 4.17 & .0579 \end{bmatrix} \begin{bmatrix} 0 & 1 & 0 \\ -1 & 0 & 0 \\ 0 & 0 & 1 \end{bmatrix} \begin{Bmatrix} .015636 \\ -.018734 \\ 1.5018 \end{Bmatrix}$$

$$f_{21} = \begin{Bmatrix} -2.7 \\ 1.2 \\ 4.5 \end{Bmatrix}$$

Check equilibrium at joint 2:

$$\Sigma M_x = 0: 1.2 - 1.2 = 0$$
$$\Sigma M_y = 0: 2.7 - 2.7 = 0$$
$$\Sigma F_z = 0: -5.5 - 4.5 + 10 = 0$$

SHEAR CENTER:

In a planar structure, if any member is loaded such that the plane of the loads does not pass through the shear center of the cross-section, the member will twist as it bends. The shear center always lies on an axis of symmetry of the cross-section. Therefore, for cases where cross-sections are used which do not contain an axis of symmetry (Fig. 7-6), the structure must be modeled so that the resulting torsion can be accounted for. In the math model, the element is usually represented by the center of gravity axis.

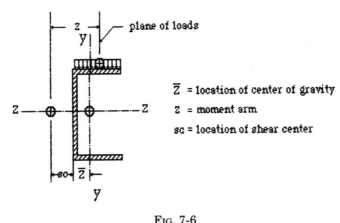

FIG. 7-6

If the center of gravity of the loads does not pass through the shear center, the load may be applied along the center of gravity axis along with a torsional moment equal to the load times the perpendicular distance from the plane of the loads to the shear center.

AXIAL LOAD ECCENTRICITY:

If the axial load in the member is not aligned with the local x axis of the member (Fig. 7-7a), the axial force causes moments about the local y and/or z axes. This is commonly the case when the member center of gravity axes do not line up with one another so that the members are connected to the joints using stiff arms (Fig. 7-7b).

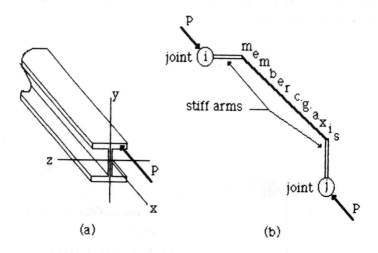

FIG. 7-7

BI-AXIAL BENDING:

While it is very common to analyse three dimensional structures using two dimensional planar models, it has been assumed, to his point, that all the members have an axis of symmetry and that the load plane is either parallel or perpendicular to that plane of symmetrey. The local x. y, z axes are assumed to represent the principal axes of the cross-section. An axis of symmetry is always a principal axis. The principal axes are always perpendicular to one another.

There are situations where the member has a symmetrical cross-section, but the plane of the loads is not parallel to one of the principle axes of the cross-section or there are in-plane loads and out-of-plane loads acting together. In those cases, the member will be subjected to bending moments about both the y and z axes simultaneously. The maximum bending stress may be doubled for a ∂ angle as small as five degrees (Fig. 7-8).

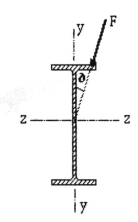

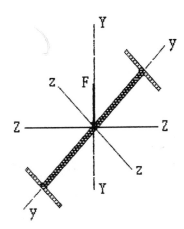

Principal Axes Parallel to Global Axes, But Load Plane is Not

Load Plane is parallel to Global Axes, But Principal Axes Are Not

(a)

(b)

Fɪɢ. 7-8

UNSYMMETRICAL BENDING:

It sometimes occurs that the cross-section does not have a plane of symmetry (Fig. 7-9). (This does not happen often for main members, but occurs frequently for secondary members and bracing.) This means that the princpal axes of the cross-section are not always parallel to the global axes. In these cases, the member is subjected to unsymmetrical bending, which means that loads which are parallel to the global axes will cause bending about the principal axes. It is the principal axes which determine the local coordinate system. The local y axis must be located relative to the global system.

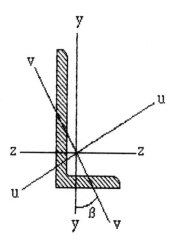

FIG. 7-9

SUPERPOSITION:

Since the only situations considered in this book deal with linear elastic behavior of structures, structures that behave as a combination of a plane frame and a grid may be analysed in two parts and the results superposed. A plane frame analysis and a grid analysis are carried out separately and the results simply added together.

$$E = 30,000 \text{ ksi}$$

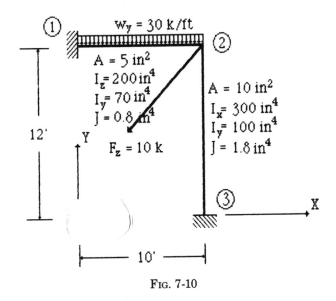

FIG. 7-10

The solution to the problem of Fig. 7-10 is the superposition of the solutions to Example Problems 6.1 and 7-1.

THREE-DIMENSIONAL STRUCTURES:

Not all structures may be adequately modeled and analysed using two dimensional models. There are times when a three dimensional analysis is warranted. This is beyond the scope of this book. However, it may be said here that the stiffness matrix representing a grid element may be combined with the stiffness matrix for a plane frame element to form the stiffness matrix for a three dimensional frame member. The stiffness matrix must be transformed in three dimensions, ie., related to the three axes of the global coordinate system.

Local Displacements

$$
\begin{array}{c}
P_x- \\
V_y- \\
V_z- \\
m_x- \\
m_y- \\
m_z-
\end{array}
k_{ii} =
\begin{bmatrix}
AE/L & 0 & 0 & 0 & 0 & 0 \\
0 & 12EI_z/L^3 & 0 & 0 & 0 & 6EI_z/L^2 \\
0 & 0 & 12EI_y/L^3 & 0 & 6EI_y/L^2 & 0 \\
0 & 0 & 0 & JG/L & 0 & 0 \\
0 & 0 & 6EI_y/L^2 & 0 & 4EI_y/L & 0 \\
0 & 6EI_z/L^2 & 0 & 0 & 0 & 4EI_z/L
\end{bmatrix}
$$

with columns labeled x, y, z, θ_x, θ_y, θ_z.

Problem 7-1: Determine the member forces in the structure shown. All loads are through the centroid of the cross sections. Both members have $A = 10$ in², $I_x = 220$ in⁴, $I_y = 80$ in⁴, $J = 1.2$ in⁴, and $E = 30,000$ ksi. Note orientation of local axes.

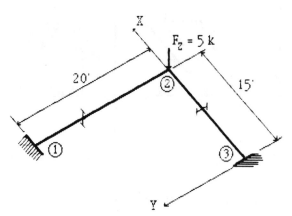

Problem 7-2: Repeat problem 7-1 with each members rotated 90° about its own longitudinal axis.

Problem 7-3: Determine the member forces in the structure shown. All loads are through the centroid of the cross sections. Ignore the fact that the beams are not in the same X-Y plane. Both members are fixed at one end and are welded together at the other end and have $A = 12$ in², $I_x = 200$ in⁴, $I_y = 50$ in⁴, $J = 1.0$ in⁴, and $E = 30,000$ ksi.

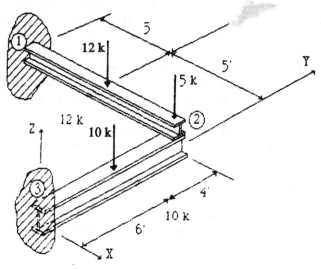

Problem 7-4: Calculate the force in the members for the structure shown. The two members are welded together at joint 2. E = 10,000 ksi, A = 5 in², I = 100 in⁴, J = 150 in⁴ for both members.

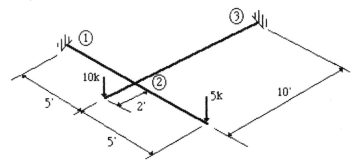

Problem 7-5 Set up the solution for the grid shown making use of symmetry.

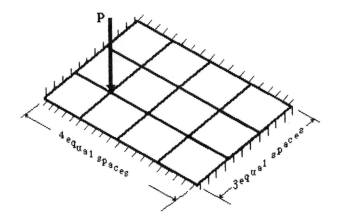

SELECTED BIBLIOGRAPHY

Carmichael, D. G., *Structural Modelling And Optimization*, New York, Wiley, 1981

Dawe, D. J., *Matrix and Finite Element Displacement Analysis of Structures*, Oxford, Clarendon Press, 1984

Desai, C. E. & Abel, J. E., *Introduction to the Finite Element Method*, New York, Van Nostrand Reinhold,1972

Grandin, H., Jr., *Fundamentals of the Finite Element Method*, New York, Macmillan, 1986

Harrison, H. B., *Computer Methods in Structural Analysis*, Englewood Cliffs, Prentice-Hall, 1973

Holzer, S. M., *Computer Analysis of Structures*, New York, Elsevier, 1985

Hughes, T. J. R., *The Finite Element Method*, Englewood Cliffs, Prentice-Hall, 1987

Kanchi, M. B., *Matrix Methods of Structural Analysis*, New York, Wiley, 1980

Kardestuncer, H., *Elementary Matrix Analysis of Structures*, New York, McGraw-Hill, 1974

McCormac, J. & Elling, R.E., *Structural Analysis*, New York, Harper & Row, 1988

McGuire, W. & Gallagher, R. H., *Matrix Structural Analysis*, New York, Wiley, 1979

Meek, J. L., *Matrix Structural Analysis*, New York, Mcgraw-Hill, 1971

Meyers, J., *Matrix Analysis of Framed Structures*, New York, Harper & Row, 1983

Mosley, W. H., *Microcomputer Applications in Structural Engineering*, New York, Elsevier, 1984

Potts, J. F., & Oler, J.W., *Finite Element Applications with Microcomputers*, New Jersey, Prentice-Hall, 1989

Przemieniecki, J. S., *Theory of Matrix Structural Analysis*, New York, McGraw-Hill, 1968

Sergerlind, L. J., *Applied Finite Element Analysis*, New York, Wiley, 1984

Shames, I. H. & Dym, C.L., *Energy and Finite Element Methods in Structural Mechanics*, New York, McGraw-Hill, 1985

Vanderbilt,M. D., *Matrix Structural Analysis*, New York, Quantum, 1974

Wang, C. K., *Structural Analysis on Microcomputers*, New York, Macmillan, 1986

Wang, C. K., *Matrix Methods of Structural Analysis*, Scranton, International Textbook,1966

Weaver, W. & Gere, J., *Matrix Analysis of Framed Structures*, 2nd Ed., New York, D. Van Nostrand, 1980

Weaver, W. Jr. & Johnston, P. R., *Finite Elements for Structural Analysis*, Englewood Cliffs, Prentice-Hall, 1984

Zienkiewicz, O.C. & Cheung, Y.K., *The Finite Element Method*, London, McGraw-Hill, 1972

Zienkiewicz, O.C., *The Finite Element Method*, 3rd Ed., London, McGraw-Hill, 1972

APPENDIX A

A SELECTED REVIEW OF MATRIX ALGEBRA

A **matrix** is a rectangular array of elements assembled in rows and columns.

The **order** of a matrix refers to its size. An **(mxn)** matrix has **m** rows and **n** columns. The **size** of a matrix is always specified row by column.

Matrices may be written in several different forms:

$$A = [a_{ij}] = \begin{bmatrix} a_{11} & a_{12} & a_{13} \\ a_{21} & a_{22} & a_{23} \\ a_{31} & a_{32} & a_{33} \end{bmatrix}$$

Where:

a_{ij} represents an element of the matrix
$i = $ row, $j = $ colum

A **Symmetrical** Matrix:

$$A = \begin{bmatrix} a & b & c \\ b & d & e \\ c & e & f \end{bmatrix}, \quad a_{ij} = a_{ji}, \text{ if } i \neq j$$

Therefore, a symmetrical matrix must be **square**, or the number of rows must equal the number of columns.

In order to **add or subtract** two matrices they must have the same number of rows and columns. The corresponding elements of each matrix are added or subtracted to form the solution matrix.

$$\begin{bmatrix} a & b \\ c & d \end{bmatrix} + \begin{bmatrix} e & f \\ g & h \end{bmatrix} = \begin{bmatrix} a+e & b+f \\ c+g & d+h \end{bmatrix}$$

To **multiply a matrix by a scalar**, each element of the matrix is multiplied by the scalar.

$$k \begin{bmatrix} a & b & c \\ b & d & e \\ c & e & f \end{bmatrix} = \begin{bmatrix} ka & kb & kc \\ kb & kd & ke \\ kc & ke & kf \end{bmatrix}$$

To perform **matrix multiplication**, the matrices must be conformable. That is; to multiply an (mxn) by a (pxq), n must equal p. The solution will be an (mxq). The product **AB**, in that order, of the (mxn) matrix $A = [a_{ij}]$ and the (nxq) matrix $B = [b_{ij}]$ is the (mxq) matrix $C = [c_{ij}]$.

$$C = AB, \qquad \overset{mxn}{A} \ \overset{nxq}{B} = \overset{mxq}{C}$$

The multiplication is carried out row by column, ie., each element of a row of matrix **A** is multiplied into the corresponding element of a column of matrix **B** and the products are summed. Each row by column multiplication is done once and only once and the sum placed in its proper place in the resulting matrix, determined by the corresponding values of (i) and (j).

$$AB = C, \quad \overset{3 \times 3}{\begin{bmatrix} a_{11} & a_{12} & a_{13} \\ a_{21} & a_{22} & a_{23} \\ a_{31} & a_{32} & a_{33} \end{bmatrix}} \overset{3 \times 2}{\begin{bmatrix} b_{11} & b_{12} \\ b_{21} & b_{22} \\ b_{31} & b_{32} \end{bmatrix}} = \overset{3 \times 2}{\begin{bmatrix} c_{11} & c_{12} \\ c_{21} & c_{22} \\ c_{31} & c_{32} \end{bmatrix}}$$

$$C_{ij} = \sum_{k=1}^{n} a_{ik}b_{kj}, \quad (i=1,2,...m; j=1,2,...q)$$

$$AB = \begin{bmatrix} a_{11}\,b_{11}+a_{12}\,b_{21}+a_{13}\,b_{31} & a_{11}\,b_{12}+a_{12}b_{22}+a_{13}\,b_{32} \\ a_{21}\,b_{11}+a_{22}\,b_{21}+a_{23}\,b_{31} & a_{21}\,b_{12}+a_{22}\,b_{22}+a_{23}\,b_{32} \\ a_{31}\,b_{11}+a_{32}\,b_{21}+a_{33}\,b_{31} & a_{31}\,b_{12}+a_{32}\,b_{22}+a_{33}\,b_{32} \end{bmatrix}$$

AB does not generally equal **BA**
AB = **AC** does not generally mean that **B** equals **C**
AB = **0** does not generally mean that **A or B** equal zero
A(BC) = **(AB)C**, associative
A(B + C) = **AB + AC**, distributive
(A + B)C = **AC + BC**, distributive

There is no **division** in matrix algebra. Instead, a matrix may be multiplied by its inverse to obtain the equivalent result.

$$\mathbf{F = Kx}, \text{ or } \mathbf{K^{-1}\,F = x}$$

The **inverse** of the matrix **A** is written as:

$$\mathbf{A^{-1}}$$

which is read **A** inverse.

A Unit Matrix:

$$\mathbf{I} = \begin{bmatrix} 1 & 0 & 0 \\ 0 & 1 & 0 \\ 0 & 0 & 1 \end{bmatrix}, \quad \begin{array}{l} a_{ij}=1, \text{ if } i=j \\[6pt] a_{ij}=0, \text{ if } i\neq j \end{array}$$

For a given square matrix $\mathbf{A}$, if another square matrix $\mathbf{B}$ of the same order as $\mathbf{A}$ can be found such that

$$AB = I$$

then $\mathbf{B} = \mathbf{A}$ inverse.

The **transpose** of a matrix is obtained by interchanging the rows and columns of the matrix.

$$\text{if } \mathbf{B} = \begin{bmatrix} a & b & c \\ d & e & f \\ g & h & i \end{bmatrix}, \text{ then } \mathbf{B^T} = \begin{bmatrix} a & d & g \\ b & e & h \\ c & f & i \end{bmatrix}$$

This is read, B transpose.

The transpose of a row matrix is a column matrix. A **row matrix** (row vector) $(1 \times n) =$. A **column matrix** (column vector) $= (n \times 1)$.

$$\begin{array}{c} (3x1) \\ \begin{bmatrix} a_{11} \\ a_{21} \\ a_{31} \end{bmatrix} \end{array} \quad \begin{array}{c} (1x3) \\ [a_{11}\ a_{12}\ a_{13}] \end{array}$$

For a symmetric matrix, $\mathbf{A} = \mathbf{A}$ transpose. If $\mathbf{A}$ is symmetric, $\mathbf{A}$ inverse is also symmetric.

An **orthogonal** matrix is a matrix whose inverse is equal to its transpose.

$$A^{-1} = A^T$$

A **minor** $(\mathbf{M_{ij}})$ of matrix $\mathbf{A}$ is formed by deleting row i and column j from the **determinate** of the matrix $\mathbf{A}$.

$$A = \begin{bmatrix} a_{11} & a_{12} & a_{13} \\ a_{21} & a_{22} & a_{23} \\ a_{31} & a_{32} & a_{33} \end{bmatrix}$$

$$M_{11} = \begin{vmatrix} a_{22} & a_{23} \\ a_{32} & a_{33} \end{vmatrix}, \; M_{12} = \begin{vmatrix} a_{21} & a_{23} \\ a_{31} & a_{33} \end{vmatrix}, \; M_{13} = \begin{vmatrix} a_{21} & a_{22} \\ a_{31} & a_{32} \end{vmatrix}, \; \text{etc.}$$

A **cofactor (Cij)** is defined as the signed minor.

$(-1)^{i+j} M_{ij}$, for example, $C_{12} = (-1)^{1+2} M_{12} = -M_{12}$

An **adjoint matrix** of a square matrix **A** is the transpose of the matrix of cofactors of **A**.

$$\text{adj } A = [C_{ij}]^T = [C_{ji}] = \begin{bmatrix} C_{11} & C_{21} & C_{31} \\ C_{12} & C_{22} & C_{32} \\ C_{13} & C_{23} & C_{33} \end{bmatrix}$$

The **inverse** of matrix **A** is the adjoint of **A** divided by the determinate of **A**.

$$[A]^{-1} = \frac{1}{A} \text{adj } [A]$$

EXAMPLE A-I:

$$\begin{Bmatrix} F_1 \\ F_2 \\ F_3 \end{Bmatrix} = \begin{bmatrix} K_{11} & K_{12} & K_{13} \\ K_{21} & K_{22} & K_{23} \\ K_{31} & K_{32} & K_{33} \end{bmatrix} \begin{Bmatrix} x_1 \\ x_2 \\ x_3 \end{Bmatrix}, \; \text{OR } F = Kx$$

$K^{-1}F = K^{-1}Kx$, $K^{-1}F = x$, Find K^{-1}

{indicates a vector, [indicates a matrix, | indicates a determinate

$$\left|K\right| = \begin{matrix} K_{11}K_{22}K_{33} + K_{12}K_{23}K_{31} + K_{13}K_{21}K_{32} - \\ K_{31}K_{22}K_{13} - K_{21}K_{12}K_{33} - K_{11}K_{32}K_{23} \end{matrix}$$

$$M_{11} = \begin{vmatrix} K_{22} & K_{23} \\ K_{32} & K_{33} \end{vmatrix} \qquad M_{23} = \begin{vmatrix} K_{11} & K_{12} \\ K_{31} & K_{32} \end{vmatrix}$$

$$M_{12} = \begin{vmatrix} K_{21} & K_{23} \\ K_{31} & K_{33} \end{vmatrix} \qquad M_{31} = \begin{vmatrix} K_{12} & K_{13} \\ K_{22} & K_{23} \end{vmatrix}$$

$$M_{13} = \begin{vmatrix} K_{21} & K_{22} \\ K_{31} & K_{32} \end{vmatrix} \qquad M_{32} = \begin{vmatrix} K_{11} & K_{12} \\ K_{21} & K_{23} \end{vmatrix}$$

$$M_{21} = \begin{vmatrix} K_{12} & K_{13} \\ K_{32} & K_{33} \end{vmatrix} \qquad M_{33} = \begin{vmatrix} K_{11} & K_{12} \\ K_{21} & K_{22} \end{vmatrix}$$

$$M_{22} = \begin{vmatrix} K_{11} & K_{13} \\ K_{31} & K_{33} \end{vmatrix}$$

$$M_{ij} = \begin{bmatrix} M_{11} & M_{12} & M_{13} \\ M_{21} & M_{22} & M_{23} \\ M_{31} & M_{32} & M_{33} \end{bmatrix}$$

$$C_{ij} = (-1)^{i+j} M_{ij} = \begin{bmatrix} M_{11} & -M_{12} & M_{13} \\ -M_{21} & M_{22} & -M_{23} \\ M_{31} & -M_{32} & M_{33} \end{bmatrix}$$

$$K^{-1} = \frac{1}{\left|K\right|} \begin{bmatrix} M_{11} & -M_{12} & M_{13} \\ -M_{21} & M_{22} & -M_{23} \\ M_{31} & -M_{32} & M_{33} \end{bmatrix}$$

EXAMPLE A-2:

Find K^{-1}:

$$K = \begin{bmatrix} 1 & 4 & 5 \\ 4 & 2 & 3 \\ 5 & 3 & 6 \end{bmatrix}$$

$$M_{11} = \begin{vmatrix} 2 & 3 \\ 3 & 6 \end{vmatrix} = 12 - 9 = 3 \qquad M_{21} = \begin{vmatrix} 4 & 5 \\ 3 & 6 \end{vmatrix} = 24 - 15 = 9$$

$$M_{12} = \begin{vmatrix} 4 & 3 \\ 5 & 6 \end{vmatrix} = 24 - 15 = 9 \qquad M_{22} = \begin{vmatrix} 1 & 5 \\ 5 & 6 \end{vmatrix} = 6 - 25 = -19$$

$$M_{13} = \begin{vmatrix} 4 & 2 \\ 5 & 3 \end{vmatrix} = 12 - 10 = 2 \qquad M_{23} = \begin{vmatrix} 1 & 4 \\ 5 & 3 \end{vmatrix} = 3 - 20 = -17$$

$$M_{31} = \begin{vmatrix} 4 & 5 \\ 2 & 3 \end{vmatrix} = 12 - 10 = 2 \qquad M_{33} = \begin{vmatrix} 1 & 4 \\ 4 & 2 \end{vmatrix} = 2 - 16 = -14$$

$$M_{32} = \begin{vmatrix} 1 & 5 \\ 4 & 3 \end{vmatrix} = 3 - 20 = -17$$

Note: $M_{ij} = M_{ji}$

$$C_{11} = (-1)^{1+1} M_{11} = 3$$
$$C_{12} = (-1)^{1+2} M_{12} = -9 = C_{21}$$
$$C_{13} = (-1)^{1+3} M_{13} = 2 = C_{31}$$
$$C_{22} = (-1)^{2+2} M_{22} = -19$$
$$C_{23} = (-1)^{2+3} M_{23} = 17 = C_{32}$$
$$C_{33} = (-1)^{3+3} M_{33} = -14$$

Note: $C_{ij} = C_{ji}$

$$\text{Cofactor of } K = \begin{bmatrix} 3 & -9 & 2 \\ -9 & -19 & 17 \\ 2 & 17 & -14 \end{bmatrix} \text{(note symmetry)}$$

$$\text{Adjoint of } K = [C_{ij}]^T = \begin{bmatrix} 3 & -9 & 2 \\ -9 & -19 & 17 \\ 2 & 17 & -14 \end{bmatrix}$$

$K^{-1} = \text{Adj } K / |K|$

$|K| = (1)(2)(6) + (4)(3)(5) + (5)(4)(3) - (5)(2)(5) - (4)(4)(6) - (1)(3)(3) = -23$

$$K^{-1} = -\frac{1}{23} \begin{bmatrix} 3 & -9 & 2 \\ -9 & -19 & 17 \\ 2 & 17 & -14 \end{bmatrix}$$

$$K^{-1} = \begin{bmatrix} -0.13 & 0.39 & -0.087 \\ 0.39 & 0.83 & -0.74 \\ -0.087 & -0.74 & -0.61 \end{bmatrix}$$

APPENDIX B

GLOSSARY

Boundary Condition or Constraint—A predetermined value of a force, displacement, and/or a temperature at the supports of a structure.

Constraint—See Boundary condition

Degree of freedom—A possible translation and/or rotation.

Displacements—Translations and/or rotations.

Element—That portion of the structure between joints.

Forces—Forces and/or moments.

Frame—A structure in which the members are rigidly connected at the joints such that shears and moments as well as axial forces may be transmitted from one member to another.

Global or Structural Coordinate System—The arbitrary, orthogonal, reference axes for the entire structure.

Grid—A structure which lies entirely in a plane with the loads acting perpendicular to that plane. Each joint may have three degrees of freedom.

Internal Forces—Forces developed internally within the structure as a result of the loads and reactions. There are member forces and joint forces.

Joint or Node—A reference point having known coordinates with respect to some arbitrarily chosen reference system.

Loads—External forces applied to a structure resulting from wind, earthquake, gravity, waves, temperature changes, differential settlement, etc.

Local Coordinate System—The orthogonal set of principal axes of each member.

Member—A discrete element of a structure.

Node—See Joint

Planar Structure—The structure and it's loads lie entirely in a plane, so that only two coordinates in space are necessary to describe the location of a point on the structure. Each node may then have up to three degrees of freedom.

Reactions—External forces at the supports resulting from the loads.

Stiff Arm—A ficticious member used to connect two joints which are rigidly connected such that there displacements are the same.

Spaceframe—A three dimensional structure requiring three coordinates in space to describe the location of a point on the structure and/or its loads. Each node may then have up to six degrees of freedom.

Spacetruss—A three dimensional structure requiring three coordinated in space to describe the location of a point on the structure and/or its loads. Each node may have up to three translational degrees of freedom.

Stiffness—Force per unit displacement

Structure—An assemblage of members and joints.

Structural Coordinate System—See Global Coordinate System

Substructure—A portion of a structure which, together with its loads and constraints, is cut from the remainder of the structure as a free body.

Truss—A structure in which all the members are pin-ended so that there are no member shears or moments, but only axial forces. Each node may have only translational degrees of freedom.

APPENDIX C

LIST OF SYMBOLS

a = angle between the local member x axis and the global X axis

a_{ij} = transformation matrix for member ij

d = displacement vector in the local coordinate system

f = local force vector

k_{ii} = member stiffness matrix, in the local coordinate system, relating forces at the i end to displacements at the i end

k_{ij} = member stiffness matrix, in the local coordinate system, relating forces at the i end to displacements at the j end

x, y, θ = local coordinates or displacements

A = cross-sectional area

C = cofactor

D = dead load

D = displacement vector in the global coordinate system

E = earthquake loads

E = modulus of elasticity

F = global force vector

G = shear modulus

I = moment of inertia

J = polar moment of inertia

J_i = stiffness matrix for joint i in the structural coordinate system

J_{ij} = stiffness matrix for member ij, at joint i, in the structural coordinate system

K = structural stiffness matrix

L = length

L = live load

L_r = roof live load

M = moment

M = minor

M_f = fixed-end moment

P = axial force

R = reactions

R = rain or ice loads

S = snow load

S_{ij} = stiffness matrix for member ij in thestructural coordinate system

V = shear force

W = weight

W = wind load

X, Y, Z = global coordinates or dislacements

$¥ = \Delta/L$

Δ = relative displacement

μ = Poisson's Ratio

r = radius of gyration

APPENDIX D

FIXED END FORCES AND MOMENTS

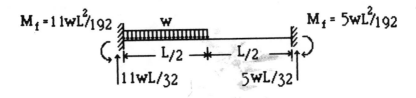

$M_f = wL^2/12$ w (Load per unit length) $M_f = wL^2/12$

L

$wL/2$ $wL/2$

$M_f = 11wL^2/192$ w $M_f = 5wL^2/192$

$\leftarrow$ L/2 $\rightarrow$ $\leftarrow$ L/2 $\rightarrow$

$11wL/32$ $5wL/32$

$M_f = PL/8$ L/2   P   L/2 $M_f = PL/8$

$P/2$ $P/2$

$M_f = Pab^2/L^2$ a   P   b $M_f = Pa^2b/L^2$

L

$Pb^2(3a+b)/L^3$ $Pa^2(3b+a)/L^3$

$M_f = 15PL/48$ L/4 P L/4 P L/4 P L/4 $M_f = 15PL/48$

$3P/2$ $3P/2$

$M_f = 2PL/9$ $M_f = 2PL/9$

$M_f = wL^2/20$ $M_f = wL^2/30$

$7wL/20$ $3wL/20$

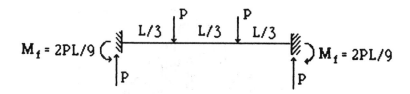

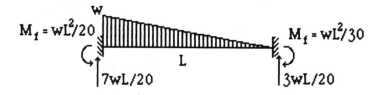

$M_f = wL^2/15$ $M_f = wL^2/15$

parabolic loading

$wL/3$ $wL/3$

$M_f = Mb(2a-b)/L^2$ a›L/3   b›L/3 $M_f = Ma(2b-a)/L^2$

$6M(aL-a^2)/L^3$ $6M(aL-a^2)/L^3$

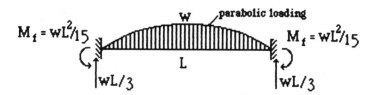

$M_f = wa^2/12L^2(3a^2-8aL + 6L^2)$ $M_f = wa^3/12L^2(4L-3a)$

$wa/2L^3(2L^3-2a^2L + a^3)$ $wa^3/2L^3(2L-a)$

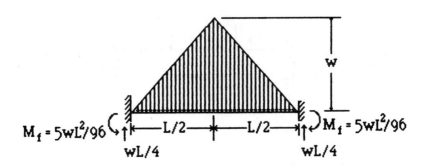

$M_f = 5wL^2/96$ $\quad\quad$ $M_f = 5wL^2/96$

$\quad\quad$ wL/4 $\quad\quad\quad\quad\quad\quad\quad\quad$ wL/4

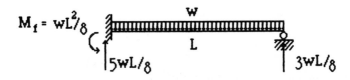

$M_f = wL^2/8$

$\quad$ 5wL/8 $\quad\quad\quad\quad\quad\quad\quad\quad$ 3wL/8

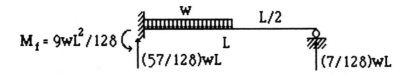

$M_f = 9wL^2/128$

$\quad$ (57/128)wL $\quad\quad\quad\quad\quad\quad$ (7/128)wL

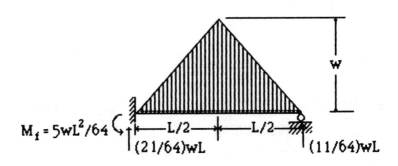

$M_f = 5wL^2/64$

$\quad$ (21/64)wL $\quad\quad\quad\quad\quad\quad$ (11/64)wL

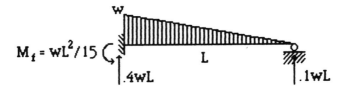

$M_f = wL^2/15$ ↺

.4wL   L   .1wL

$M_f = 3PL/16$ ↺

L/2   P   L/2

11P/16   5P/16

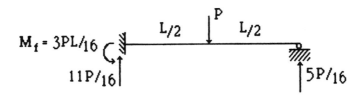

$M_f = (P/L^2)[b^2a + a^2b/2]$ ↺

a   P   b

L

$R_A = P - R_B$ $R_B = (P/L)[b^2a + a^2b/2 - a]$

$M_f = PL/3$

P   P

L/3   L/3   L/3

4PL/3   2PL/3

$M_f = 45PL/96$ ↺

P   P   P

L/4   L/4   L/4   L/4

(63/32)P   (33/32)P